Made to Break

GILES SLADE

Harvard University Press

Cambridge, Massachusetts | London, England

Made To Break

Technology and Obsolescence in America

First Harvard University Press paperback edition, 2007

Library of Congress Cataloging-in-Publication Data

Slade, Giles.
    Made to break : technology and obsolescence in America /
    Giles Slade.
        p. cm.
    ISBN-13 978-0-674-02203-4 (cloth: alk. paper)
    ISBN-10 0-674-02203-3 (cloth: alk. paper)
    ISBN-13 978-0-674-02572-1 (pbk.)
    ISBN-10 0-674-02572-5 (pbk.)
    1. Technological innovations—United States.   I. Title.
    T173.8.S595 2006
    609.73—dc22   2005036315

Contents

America, I do not call your name without hope

—PABLO NERUDA

To scrutinize the trivial can be to discover the monumental. Almost any object can serve to unveil the mysteries of engineering and its relation to art, business, and all other aspects of our culture.

HENRY PETROSKI, *THE PENCIL: A HISTORY* (1989)

Introduction

For no better reason than that a century of advertising has conditioned us to want more, better, and faster from any consumer good we purchase, in 2004 about 315 million working PCs were retired in North America. Of these, as many as 10 percent would be refurbished and reused, but most would go straight to the trash heap. These still-functioning but obsolete computers represented an enormous increase over the 63 million working PCs dumped into American landfills in 2003. In 1997, although a PC monitor lasted six or seven years, a CPU was expected to last only four or five. By 2003 informed consumers expected only two years of use from the new systems they were purchasing, and today the life expectancy of most PCs is even less.[1]

In 2005 more than 100 million cell phones were discarded in the United States. This 50,000 tons of still-usable equipment joined another 200,000 tons of cell phones already awaiting dismantling and disposal. Unlike PCs, the compact design of cell phones resists disassembly for recycling—it's much easier just to throw phones away and make new ones. So despite the fact that they weigh only a fraction of what PCs weigh, discarded cell phones represent a toxic time bomb waiting to enter America's landfills and water table.[2]

Cell phones and PCs travel in the company of a vast assortment of obsolete IT electronics, including last year's Palms, Blackberries, Notebooks, printers, copiers, monitors, scanners, modems, hubs, docking ports, digital cameras, LCD projectors, Zip drives, speakers, keyboards, mice, GameBoys, Walkmen, CD players, VCRs, and DVD players—all awaiting disposal. PlayStations, Xboxes, and iPods are not far behind. Obsolete cathode ray tubes used in computer monitors will already be in the trash (superseded by LCDs, as in Japan) by the time a U.S. government mandate goes into effect in 2009 committing all of the country to High-Definition TV. The CRTs of analog televisions are constructed along the same general design as those of PC monitors, but they are larger—often *much* larger—and are made up of about 55 percent toxic lead glass, while a monitor is only about 28 to 36 percent. But the looming problem is not just the oversized analog TV sitting in the family room, which will require a team of professional movers to haul away. The fact is that no one really knows how many smaller analog TVs still lurk in basements, attics, garages, and kitchens, not to mention the back rooms of sports bars, fitness clubs, and other commercial sites.

What *is* known is frightening. Since the 1970s, TV sales have achieved about a 95 percent penetration rate in American homes, compared to the 50 percent penetration rate computers achieved in the 1990s. For more than a decade, about 20 to 25 million TVs have been sold annually in the United States, while only 20,000 are recycled each year. So as federal regulations mandating HDTV come into effect in 2009, an unknown but substantially larger number of analog TVs will join the hundreds of millions of computer monitors entering America's overcrowded, pre-toxic waste stream. Just this one-time disposal of "brown goods" will, alone, more than double the hazardous waste problem in North America.[3]

Meanwhile, no one has figured out what to do with plain old telephone service receivers, whose lead-solder connections and PVC cases are quickly becoming obsolete as consumers make the switch to 3G cell phones and VoI (voice over the Internet). As these archaic devices are piled on top of other remnants of wired technology, America's landfills—already overflowing—will reach a point where they can no longer offer a suitable burial for the nation's electronic junk.[4]

Until recently the United States shipped much of its toxic e-waste to China, India, Pakistan, Bangladesh, and other economically desperate countries in the developing world. But exportation is, at best, a stop-gap strategy. Following the Basel Convention, the United Nations slowed electronic waste shipments to these ports. But more practically, the e-waste problem will soon reach such gigantic proportions that it will overwhelm our shipping capacity. The world simply cannot produce enough *containers* for America to continue at its current level as an exporter of both electronic goods and electronic waste. Consequently, all of these discarded and highly toxic components represent an insurmountable future storage problem. We do not have enough time, money, or space in the continental United States to create enough landfills to store and then ignore America's growing pile of electronic trash.[5]

What brought us to this pass?

DELIBERATE OBSOLESCENCE IN ALL ITS FORMS—technological, psychological, or planned—is a uniquely American invention.

Not only did we invent disposable products, ranging from diapers to cameras to contact lenses, but we invented the very concept of disposability itself, as a necessary precursor to our rejection of tradition and our promotion of progress and change. As American manufacturers learned how to exploit obsolescence, American consumers increasingly accepted it in every aspect of their lives. Actual use of the word "obsolescence" to describe out-of-date consumer products began to show up in the early twentieth century when modern household appliances replaced older stoves and fireplaces, and steel pots replaced iron ones. But it was the electric starter in automobiles, introduced in 1913, that raised obsolescence to national prominence by rendering all previous cars obsolete. Even the most modern American women hated hand-cranking their cars and were greatly relieved when they could simply push a start button on a newer model.[6] The earliest phase of product obsolescence, then, is called *technological obsolescence,* or obsolescence due to technological innovation.

The second stage of product obsolescence occurred about a decade later, in 1923. Executives who had migrated to General Motors from the chemical and dye-making giant DuPont adapted a marketing strategy from what was then America's third largest and most rapidly growing industry: textiles and fashions. Instead of waiting for technological innovations that would push consumers to trade in their older-model cars, General Motors turned to sleek styling as a way of making newer cars more desirable and pulling potential buyers into the showroom. The success of GM's cosmetic changes to the 1923 Chevrolet indicated that consumers were willing to trade up for style, not just for technological improvements, long before their old cars wore out. This strategy was so successful that it spread quickly to many other American industries, such as watches and radios. The annual model change adopted by carmakers is an example of *psychological, progressive,* or *dynamic obsolescence.* All of these terms refer to the mechanism of changing product style as a way to manipulate consumers into repetitive buying.

The most recent stage in the history of product obsolescence began when producers recognized their ability to manipulate the failure rate of manufactured materials. After prolonged use, any product will fail because its materials become worn or stressed. This is normal. But during the Depression, manufacturers were forced to return to the practice of adulteration—the nineteenth-century technique of using inferior materials in manufactured goods—as a simple cost-cutting measure: inferior materials lowered unit costs. But these same manufacturers soon realized that adulteration also stimulated demand. After a decade of unprecedented affluence and consumption during the 1920s, consumer demand fell radically with the onset of the Depression, and in desperation manufacturers used inferior materials to deliberately shorten the life spans of products and force consumers to purchase replacements.

Planned obsolescence is the catch-all phrase used to describe the assortment of techniques used to artificially limit the durability of a manufactured good in order to stimulate repetitive consumption. To achieve shorter product lives and sell more goods, manufacturers in the 1930s began to base their choice of materials on scientific tests by newly formed research and development departments. These tests determined when each of the product's specific components would fail. One of the few known examples of this monopolistic (and hence illegal) strategy was a change, proposed but never implemented, to shorten the life of General Electric's flashlight bulbs in order to increase demand by as much as 60 percent.

As obsolescence became an increasingly useful manufacturing and marketing tool, an eclectic assortment of advertisers, bankers, business analysts, communications theorists, economists, engineers, industrial designers, and even real estate brokers contrived ways to describe, control, promote, and exploit the market demand that obsolescence created. What these approaches had in common was their focus on a radical break with tradition in order to deliver products, and prosperity, to the greatest number of people—and in the process to gain market share and make a buck. Both goals strike us today as quintessentially American in spirit.

But even as these professionals were inventing the means to exploit obsolescence, a number of articulate American critics began to see this manipulation of the public as the very epitome of what was wrong with our culture and its economic system. The former journalist Vance Packard raised the issue powerfully in his debut book, *The Hidden Persuaders,* in 1957, which revealed how advertisers relied on motivational research to manipulate potential buyers. Others, including Norman Cousins, John Kenneth Galbraith, Marshall McLuhan, Archibald MacLeish, and Victor Papanek, would follow Packard's lead in pointing out how the media create artificial needs within vulnerable consumers. The sheer volume of print Americans have devoted to this topic since 1927 demonstrates that obsolescence has become a touchstone of the American consciousness.

The book you have in your hand is a collection of stories that emerged during my search for obsolescence in uniquely American events: the invention of packaging, advertising, and branding; the rivalry between Ford and GM; "death dating"; the invention of radio, television, and transistors; the war and the postwar competition with Japan; rock and roll, the British Invasion, and male fashions; universal home ownership; calculators, integrated circuits, and PCs; the space race, tailfins, and TelStar; and the looming cri-sis of e-waste. The theory and practice of obsolescence play a cen-tral role in each of these American milestones. At each juncture, vested interests struggled and competed to achieve repetitive con-sumption through obsolescence, in its many forms and combina-tions.

A FEW YEARS BACK, as I was visiting a touring exhibit called "Eternal Egypt" with my ten-year-old son, it occurred to me that while the ancient Egyptians built great monuments to endure for countless generations, just about everything we produce in North America is made to break. If human history reserves a privileged place for the Egyptians because of their rich conception of the afterlife, what place will it reserve for a people who, in their seem-ing worship of convenience and greed, left behind mountains of electronic debris? What can be said of a culture whose legacies to the future are mounds of hazardous materials and a poisoned water supply? Will America's pyramids be pyramids of waste? The point of this book is to raise this troubling question.

A few foresaw a [world] . . . in which the ever-expanding taste for material goods and the theory of comparative advantage would keep us all running as fast as we could on a giant squirrel wheel.

JAMES KATZ, *MACHINES THAT BECOME US* (2003)

1   Repetitive Consumption

Long before mass production became a universally accepted term in the 1950s, American businessmen worried about overproduction and how to avoid it—not by producing less but by selling more. As the late nineteenth-century economy changed from man-powered to machine-driven industry, manufacturers became painfully aware that their factories could now produce more goods than could be readily distributed and consumed. America was "suffering from overproduction," a frustrated retailer wrote in 1876. "The warehouses of the world are filled with goods."[1]

Half a century later, the inventor of disposable razors, King Camp Gillette, still considered overproduction to be America's most troubling social evil: "We have the paradox of idle men, only too anxious for work, and idle plants in perfect conditions for production, at the same time that people are starving and frozen. The reason is overproduction. It seems a bit absurd that when we have overproduced we should go without. One would think that overproduction would warrant a furious holiday and a riot of feasting and display of all the superfluous goods lying around. On the contrary, overproduction produces want."[2]

As American manufacturers and retailers thought about solutions to this industrial-age dilemma, they decided that the problem of overproduction was twofold. The first problem was *demand*—how to create it and how to sustain it. The related problem was *distribution*—how to move goods swiftly and profitably from factories to consumers.[3] From the late nineteenth century onward, Americans confronted the problem of distribution head-on, through the development of national highways, cheap and reliable railroad freight, mail-order houses such as Sears, Roebuck and Montgomery Ward, department stores such as Bloomingdale's, Wanamakers, and Marshall Field, and eventually national retail chains like Macy's and, in recent times, the merchandising giant Wal-Mart.

While retailers were developing a national distribution network, manufacturers attacked the problem of slack demand by developing innovative marketing campaigns. Advertising would play a major role here, but what was it about their goods that manufacturers should advertise? Before consumer ads could become effective in creating a demand for a product, the product had to be differentiated in some way from similar goods. Why was Uneeda Biscuit preferred over Iwanna Biscuit? The goal was not simply to increase biscuit consumption per se but to create repetitive consumption of one's own brand, which would relieve overproduction. The central marketing question of the early twentieth century was how could a manufacturer encourage consumers to return to his product again and again, instead of buying the wares of his competitor?

Solutions to the problem of how to promote repetitive consumption would eventually include a wide range of manufacturing strategies, from branding, packaging, and creating disposable products to continuously changing the styles of nondisposable products so that they became psychologically obsolete. All such strategies derive from a marketing question first expressed in the late nineteenth century but most succinctly rendered in 1925 by Edward Filene, the influential Boston department store magnate: "How can I manage my business . . . so that I can be sure of a permanent and growing body of consumers?"[4]

BRANDING AND PACKAGING

The first answer that manufacturers found was branding. In the 1850s a handful of products, including Singer sewing machines and McCormick agricultural machinery, began to display the company name prominently, as the initial step in establishing a direct relationship between the company and its customers. Singer also provided financing, service, trade-ins, and authorized local dealers who educated clients in the use and maintenance of their expensive machines, eliminating shopkeepers as middlemen.[5]

Branding soon became closely associated with another strategy for creating repetitive demand: packaging. Manufacturers of foodstuffs could not screw a metal nameplate onto their products, but they could advertise their brand by enclosing those products in fancy packaging. As a practical matter, individual packaging allowed manufacturers to distribute their product more widely. And in a few cases, modern packaging itself became the focus of a successful advertising campaign. In 1899 the National Biscuit Company, makers of Uneeda Biscuit, began to feature its patented In-Er-Seal prominently in a national campaign to create demand for their product. Before Nabisco developed this new marketing strategy, consumers had bought biscuits (also called crackers) in bulk from an open cracker barrel in a local store. At a time when bakeries were scarce and crackers were a more common staple than bread, National Biscuit emphasized that Uneeda's In-Er-Seal package prevented moisture from ruining the quality and flavor of their biscuits. They supported this campaign with a wonderful newspaper and handbill graphic that depicted a boy in a yellow slicker pushing a wheelbarrow full of biscuit boxes home in the rain. Eventually the boy in a yellow slicker became ubiquitous on Uneeda packaging, and customers asked for Uneeda biscuits by name. The National Biscuit Company had successfully created enough demand for its product to guarantee repetitive consumption and to free Nabisco from problem of overproduction.

Three other companies—Wrigley's, the American Tobacco Company, and Procter & Gamble—adopted similar tactics to establish product loyalty among their customers even before the historic Uneeda campaign. They designed strong national ad programs not just to identify their brands but to provide reassuring guarantees of quality. Such guarantees were necessary for customers who bought most staples in bulk and were suspicious of any packaging that prevented them from testing, tasting, or sampling. In time, as promotional campaigns became more sophisticated, consumers overcame their qualms about packaged goods. They realized that a piece of Juicy Fruit gum, a box of United Cigars, or a bar of Ivory Soap would always be the same, no matter where it was bought. Modern packaging, with its trademarks and identifying logos, guaranteed that those products would be of consistent quality and safe to buy. And because these products were distributed nationally, brand names assured consumers of an equitable value-for-money exchange at any store in the country.

The home efficiency expert Christine Frederick observed in 1919 that "the one means of protection the consumer can rely on is the 'trademark' on the package or product she buys . . . In every case, the trademarked brand carries more integrity or guarantee."[6] By the turn of the century Americans were getting into the comfortable habit of remembering their favorite brands and asking for them by name.

DISPOSABLE PRODUCTS FOR MEN

Manufacturers developed other strategies besides branding to encourage repetitive consumption. What has been called "disposable culture" or "the throwaway ethic" began in America around the middle of the nineteenth century when a variety of cheap materials became available to industry. Innovations in the machinery of paper production, for example, made paper a practical substitute for cloth. The millions of paper shirt fronts (bosoms, as they were then called), as well as the collars and cuffs that adorned nineteenth-century American men, owe their commercial success to this technological advance.

The beauty of these disposable products, as far as paper manufacturers were concerned, was that demand for them seemed endless. In 1872 America produced 150 million disposable shirt collars and cuffs. Men found paper clothing parts convenient because laundry services in those days were unreliable, expensive, and available mainly in large urban centers. America was still predominantly a rural culture, and before the advent of modern washing machines in the twentieth century, laundry was an onerous, labor-intensive task undertaken by women once weekly on Blue Tuesday. Single men simply lacked access to professional or spousal laundry services. They bought replaceable shirt parts in bulk and changed into them whenever the most visible parts of their attire became stained or discolored. Disposing of a soiled cuff, collar, or bosom was as easy as dropping it into the nearest fireplace or pot-bellied stove.

At the same time that paper was becoming cheaper and more plentiful, America experienced a revolution in steel production. Two watchmakers, Waterbury and then Ingersoll, took advantage of this industrial development to roll out a national distribution plan for low-priced, highly reliable steel watches. In the 1880s Waterbury produced pocket watches at such low cost that they became promotional giveaways to department store customers who bought expensive winter coats. When Reginald Belfield entered George Westinghouse's employ in 1885, he spruced up his image with a ready-made coat and complimentary Waterbury watch from Kaufmann's in Pittsburgh. In a testament to its standard of manufacture, Belfield relates that Westinghouse was fascinated with the watch. He "took it to pieces many times and put it together again . . . The watch never suffered for the treatment."[7]

Ten years later, as standards of watch manufacture became more relaxed, Ingersoll introduced the Yankee and billed it as "the watch that made the dollar famous," since it sold for exactly a buck. Like Ingersoll's mail-order dollar-and-a-half watch before it, the Yankee kept accurate time for at least a year. Its reliability and insignificant cost—other watch prices hovered around the $10 mark in the early 1890s, when the average wage was around a dollar a day—made the Yankee immediately popular.[8]

The low price had another important effect on Ingersoll customers. Despite a reliable mail-in guarantee offering free replacements whenever an Ingersoll watch went wrong, working-class Yankee owners simply threw their failing watches away and bought a replacement. During the entire period of its production (from 1901 to 1914), only 3 percent of Yankee watches were ever returned for replacements. At around the same time that the phrase "instant gratification" first began to appear in popular magazines, Yankee owners explained that their need for a time-piece was greater than their need to save a dollar by waiting several weeks for a free replacement to arrive by return mail.

Before very long, the venture into planned obsolescence by these two innovative American watchmakers ended with the un-planned obsolescence of their own dollar pocket watch. As wrist-watches came into fashion near the end of World War I, pocket watches became obsolete. By the time Waterbury was sold to a group of Scandinavian businessmen to produce fuse timers for war ordnance in 1942, the company was close to bankruptcy.[9]

The same revolution in steel production that had produced cheap steel for Ingersoll encouraged other disposable products. In the 1890s William Painter, a Baltimore dye maker, invented and patented an inexpensive cork-lined replacement for rubber bottle stoppers called the Crown Cork (or Crown Bottle Cap). Famous today among a growing body of collectors who specialize in early disposables, Painter's earliest caps are distinctive because they have only 21 crimpings as opposed to the later standard of 24. The bottle cap itself was made of pressed tinplate, a metal that had been used in throwaway cans since 1813 and in cheap children's toys since mid-century. But Painter is most significant in the history of planned obsolescence because he hired, befriended, and then encouraged the 36-year-old salesman King Camp Gillette to invent products which, like the bottle cap, were used only once and then tossed in the trash. Painter himself never used the terms repetitive consumption or planned obsolescence, but the implications of his advice are unmistakable. "Think of something like the Crown Cork," he told Gillette one evening in the extravagant parlor of his opulent Baltimore mansion. Once it is used and thrown away, "the customer keeps coming back for more."[10]

The tale of how King Gillette invented his disposable razor blade reached legendary status in early twentieth-century America—much as the story of Microsoft's founding or the origin of Apple computers is widely known today. By the time Sinclair Lewis's *Babbitt* was published in 1922, most Americans knew that while King Gillette was shaving with a Star safety razor in Pennsylvania one morning in 1895, he realized that his blunt steel blade needed to be professionally resharpened. "I found my razor dull, and not only was it dull but it was beyond the point of successful stropping and it needed honing for which it must be taken to a barber or a cutler. As I stood there with the razor in my hand, my eyes resting on it as lightly as a bird settling down on its nest— the Gillette razor was born." As King thought about the high maintenance required of Star blades, not to mention their high cost—up to $1.50 apiece, a significant amount of money in the 1890s—he must have wondered: what if a cheaper, thinner blade could be stamped out of sheet metal and then honed on two edges? When both sides become dull, the customer can simply replace it with a new one. That very day, Gillette wrote to his wife, Alanta, "Our fortune is made," and he was almost right.[11]

Another six years would pass before a process could be developed that would interleave sheet steel and copper in order to allow thin metal sheets to temper without buckling. Only thin metal so tempered could hold the razor-sharp edge needed for Gillette's disposable blades. By 1905, using the slogan "No stropping. No Honing," Gillette's safety razor had won public acceptance and begun its steep trajectory of growth. Like paper shirt parts and Ingersoll watches before it, this disposable product targeted men. It also circumvented accusations of unthrifty wastefulness because it was convenient and hygienic, elements of the story that Sinclair Lewis later chose to satirize in 1922.

One morning, Lewis wrote, George F. Babbitt was confronted with a disposable razor blade just as dull as Gillette's Star blade in 1895: "He hunted through the medicine-chest for a packet of new razor blades (reflecting, as invariably, 'Be cheaper to buy one of those dinguses and strop your own blades,') and when he discovered the packet, behind the round box of bicarbonate of soda, he thought ill of his wife for putting it there and very well of himself for not saying 'Damn'. But he did say it, immediately afterward, when with wet and soap-slippery fingers he tried to remove the horrible little envelope and crisp clinging oiled paper from the new blade . . . Then there was the problem . . . of what to do with the old blade, which might imperil the fingers of his young. As usual, he tossed it on top of the medicine-cabinet, with a mental note that someday he must remove the 50 or 60 other blades that were . . . temporarily piled up there."[12]

In addition to revolutions in paper and steel production that made male-oriented disposable products possible, a little known but equally revolutionary process in rubber manufacture in the middle of the nineteenth century demonstrates the early connections among disposability, repetitive consumption, hygiene, and health. Long before Sarah Chase and Margaret Sanger began their enlightened distribution of birth control devices to women in defiance of the repressive Comstock Laws, the development of vulcanization in the 1850s made possible the manufacture of a heavy style of condom from rubber. Vulcanized rubber condoms were not very elastic, however, and so they were fastened in place by two attached ribbons. Gradually, as these contraceptive devices gained popularity, they came to be known as "rubbers."

Rubbers were distinct from another kind of condom called

a "skin," which was made from sheep intestine. Because skins were more difficult to manufacture, they were (and remain today) much more expensive than rubber condoms. For this reason, nineteenth-century bordello owners sometimes collected skins after use, laundered, dried, and recirculated them. The coming of vulcanized rubber condoms changed all that. Rubbers were much cheaper than skins. After "tying one on" (a male expression that eventually became associated with the drinking party preceding intercourse), a man disposed of his used rubber. Each additional sexual encounter required a new purchase. After the development of latex condoms in the 1880s, rubbers became more elastic. They kept the same name, but (to the dismay of some traditionalists) they lost their pretty ribbons.

DISPOSABLE PRODUCTS FOR WOMEN

In the early decades of the twentieth century, manufacturers who had embraced disposability as a viable way to achieve repetitive consumption realized that in catering mainly to men, they severely limited their potential market. Urbanization and industrialization had changed American gender roles, and single women were entering the workforce in greater numbers. Brochures published by the Women's Educational and Industrial Union of Boston record the variety of employment opportunities available to women. Many of the suggested positions were previously restricted to men. They included work in publishing, real estate, probation, industrial chemistry, and bacteriology.[13]

Changes in laws concerning inheritance and the integrity of life insurance policies, and especially improvements in their enforcement, were also putting more money into widowed women's hands.[14] Furthermore, America had shifted from a subsistence agrarian economy to an industrial one, and as a result more and more married women found themselves in cities, shopping for their families' needs in the hours that their husbands worked and their children went to school. There was much more money to go around, and many more things to buy. Women were suddenly disproportionately in charge of spending. By the 1920s some writers of the time claimed that women made 85 percent of all consumer purchases, including automobiles and men's fashions.

As early as 1907, an adman recognized women's influence over the family budget. W. H. Black, manager of *The Delineator,* an early advertising trade magazine, wrote in that year that women "are the spenders of [the family] income. Upon them is the larger responsibility of obtaining one hundred cents worth of value for the dollar spent."[15] Among the first businesses to capitalize on these social and demographic changes were advertisers and publishers. By the beginning of World War I some advertisers began to hire women as copywriters to manage accounts targeted at female consumers. From 1916 on, J. Walter Thompson, the oldest advertising agency in the world, established a Women's Editorial Department at their New York offices to handle the account for Cutex, the celluloid nail varnish company. This unique group included many socially active, liberal feminists from prominent backgrounds.[16] Among the JWT staff was Frances Maule, an organizer for the New York State Suffrage Party and a speaker for the National Women Suffrage Association. Maule soon became one of JWT's senior copywriters. By 1918 the Women's Editorial Department was responsible for nearly 60 percent of J. Walter Thompson's annual business, $2,264,759 in billings.

The following year, Congress enfranchised women by passing the Nineteenth Amendment, and Mary Macfadden, wife of the publisher of *Physical Culture Magazine,* founded *True Story.*

The immediate popularity and subsequent rise in monthly sales of *True Story* indicate the growing purchasing power of blue-collar women. Each *True Story* issue cost a whopping 20 cents, compared to 15 cents for the *Ladies' Home Journal*. And like the *LHJ*, *True Story* was sold only at newsstands. Nevertheless, by 1924 around 850,000 copies of *True Story* were bought every month; by 1927, with monthly sales exceeding 2 million copies, it was more popular than *Ladies' Home Journal* or *McCall's*. Women's magazines generally, and the confessional magazine in particular, had become a medium that advertisers could exploit to sell all kinds of personal products to women.[17]

Shortly after *True Story* hit the newsstands for the first time, manufacturers began developing personal products specifically designed to encourage repetitive consumption by women. In 1916 a Kimberly-Clark subsidiary had stockpiled a new absorbent material made from celluloid. The Cellucotton company's single product was originally intended for use in military bandages and gas mask filters, since the war in Europe was expected to last at least eight years. When World War I unexpectedly ended in 1918, Kimberly-Clark was saddled with a formidable supply of their new material. Confronted with this costly surplus, they did some fast thinking. Fourteen months later, in 1920, Kimberly-Clark introduced a disposable sanitary napkin called Kotex.[18]

Montgomery Ward had first manufactured sanitary napkins in 1895. They were designed for a single-use convenience, but the early versions were expensive, and the products they replaced were either scraps from a woman's ragbag or an inexpensive textured cotton called Birdseye. Since Kotex was made of war-surplus material, it could be manufactured and sold cheaply. At a nickel each, sanitary napkins fit the budgets of most urban women. A nickel was also the coin of choice in nickelodeons, automats, and vending machines—a point not lost on Kimberly-Clark, which made Kotex immediately available to women in need, by installing vending machines in ladies' rooms across the country.

Kimberly-Clark hired Charles F. Nichols, a Chicago-based ad agency, to develop a Kotex ad campaign tailored to women's magazines that would depict women outside their traditional role as homemaker. In the earliest Kotex ad, a woman in an elegant ball gown descended a spiral staircase, seemingly unconcerned about "that time of the month." "Today, with Kotex," the ad promised, "you need never lose a single precious hour." Later ads showed women skating and traveling—their freedom, mobility, and independence made possible by disposable sanitary napkins.[19]

Despite the national success and continued growth of its Kotex brand, Kimberly-Clark changed advertising agencies in 1924, giving the account to Albert Lasker, head of Lord and Thomas. Lasker's explanation for why he deliberately pursued the Kotex account underlines the market potential of disposable products: "The products I like to advertise most," he said later, "are those that are only used once."[20] Lasker was also aware that the Kimberly-Clark account might soon double in size. Late in 1924 the company debuted a second disposable product for women made of the same war-surplus cellucotton as Kotex. (They still had plenty on hand.) This new item was called Kleenex—a brand name so popular that it eventually became the generic term used for disposable tissues. Kleenex was originally marketed as a make-up removing, face-cleansing product for women. These "disposable kerchiefs," as they were called, offered women a new way to remove cold cream. Soon, however, American women began using disposable tissues to blow their noses. By 1927 Kimberly-Clark had picked up on this trend and had changed their advertising accordingly, now recommending Kleenex for sanitary purposes during the cold and flu season.

Other manufacturers soon noticed the popularity of disposable hygienic products for women. In 1921 Johnson & Johnson introduced the first hand-made Band-Aids (another brand name that came into generic use). Originally invented in 1920 as a quick way to minister to minor cuts on fingers that occurred during food preparation, Band-Aids eliminated the difficulty of having to construct a bandage from cotton and tape while one's fingers were bleeding. But despite their convenience, Band-Aids did not become popular until Johnson & Johnson changed its method of manufacture. In 1924 the company switched from hand-made to machine-made smaller Band-Aids, which could be produced in greater numbers and with the guarantee that the product was uniform and sterile. This made a world of difference in Band-Aid's success.

Johnson & Johnson lost no time in developing other disposable products for women. In 1926, the same year she became the first woman member of the American Society of Mechanical Engineers, Lillian Gilbreth researched one of these new products. Trained in psychology as well as ergonomic design for women, Gilbreth was uniquely qualified to tell Johnson & Johnson what they wanted to know: how to design a sanitary napkin that was smaller, thinner, and more comfortable than Kotex, the leading brand. In 1927, following Gilbreth's recommendations, Johnson & Johnson introduced a form-fitted sanitary napkin that was much more readily disposable than any other available product. Called Modess, this streamlined highly absorbent sanitary napkin introduced real competition to a market Kotex had previously dominated. The resulting loss in market share sent Kimberly-Clark back to the drawing board. It unveiled its own new Phantom Kotex in 1932.

Clearly, women were getting the disposable habit. A growing demand for the still-shrinking sanitary napkin led, in 1934, to the marketing of tampons, a commercial product *Consumer Magazine* later recognized as among the fifty most revolutionary products of the twentieth century. Designed by Dr. Earle Haas of Denver and produced by a company first owned and operated by Gertrude Tenderich, Tampax, along with other disposables, not only habituated women to increasing levels of repetitive consumption but broadened the cultural acceptance of the throwaway ethic, a necessary accompaniment to planned obsolescence. Not only were tampons and sanitary napkins tossed in the trash after one use, but such products also gave more affluent women one less reason to hoard scraps of cloth, as their forebears had done. Among this monied group, the ready-to-wear fashions of the day could be quickly disposed of and replaced once they were no longer in style.

Throughout America, old garments were thrown away as never before. There was less reason to save rags and more stigma attached to doing so. The pejorative expression "on the rag" dates from this period, when advertising for sanitary napkins lifted some of the social taboos surrounding menstruation and allowed for more direct expression. Earlier slang phrases had been much more coded and obscure (falling off the roof, visiting Auntie, waving a flag, wearing red shoes, too wet to plow, and the simple but ubiquitous "curse").

ANTI-THRIFT CAMPAIGNS

Encouraged by the repetitive consumption of disposable paper products for both men and women, paper manufacturers developed toilet paper, paper cups, paper towels, and paper straws (rendering rye stalks obsolete). Gradually, the popularity of disposable personal products, purchased and used in the name of hygiene and health, caused Americans to generalize their throwaway habit to other goods.[21] This was a significant development in the history of product obsolescence. As a throwaway culture emerged, the ethic of durability, of thrift, of what the consumer historian Susan Strasser calls "the stewardship of objects," was slowly modified. At first, people just threw their paper products into the fire. But as the disposable trend continued, it became culturally permissible to throw away objects that could not simply and conveniently be consumed by flames.

Americans displayed a profound ambivalence concerning thrift and waste, going back at least as far as the late nineteenth century. The widespread encouragement of domestic thrift by home economists such as Catharine Beecher and the ethic of durability championed by the Craftsman movement coexisted with massive public wastefulness. This extravagance is eloquently described in the memoir of a young Dutch immigrant who would later edit the *Ladies' Home Journal.* In a memoir published in 1921, Edward Bok recalls his impressions on arriving in America in the late 1870s:

> We had been in the United States only a few days before the realization came home strongly to my father and mother that they had brought their children to a land of waste . . . There was waste, and the most prodigal waste, on every hand. In every streetcar and on every ferryboat the floors and seats were littered with newspapers that had been read and thrown away or left behind. If I went to a grocery store to buy a peck of potatoes, and a potato rolled off the heaping measure, the grocery man, instead of picking it up, kicked it into the gutter for the wheels of his wagon to run over. The butcher's waste filled my mother's soul with dismay. If I bought a scuttle of coal at the corner grocery, the coal

that missed the scuttle, instead of being shoveled up and put back into the bin, was swept into the street. My young eyes quickly saw this; in the evening I gathered up the coal thus swept away, and during the course of a week I collected a scuttle full . . . At school, I quickly learned that to *save money* was to be *stingy;* as a young man, I soon found that the American disliked the word "economy," and on every hand as plenty grew spending grew. There was literally nothing in American life to teach me thrift or economy; everything to teach me to spend and to waste.[22]

On the other hand, the thrift advocated by American home economists was sometimes taken to ridiculous extremes in American public life, and, not surprisingly, it provoked a backlash. During Taft's presidency, for example, Frank H. Hitchcock, postmaster general of the United States, ordered his clerks to extend the life of their characteristic anidine pencils or "reds" by issuing them with a tin ferrule that extended the length and life of a pencil stub. Before long, Americans jeeringly referred to all pencils as Hitchcocks. Disposable products, throwaway packaging, and changing fashions were making it more and more acceptable to be unthrifty by discarding whatever was not immediately useful.[23]

During World War I, the national frugality campaign organized by Treasury Department appointee Frank Vanderlip met substantial resistance. Merchants across America rejected Vanderlip's encouragement of thrift because it threatened to ruin their Christmas business. Late in 1917, stores in every city began displaying signs reading "Business as Usual. Beware of Thrift and Unwise Economy." Local newspapers weighed in, supporting their advertisers, the retailers. Editorials championed "Business as Usual" across the nation well into 1918, and Boston papers refused to run a series of patriotic ads supporting the thrift campaign sponsored by an assortment of local academics.[24]

By the time the war ended in November 1918, thrift was a dying issue. It revived temporarily in 1920 when a sudden depression withered retail sales nationally. But Henry Ford's ghostwriter, Samuel D. Crowther, expressed the nation's strongest feelings about thrift in 1922 when he wrote: "What can be fine about paring the necessities of life to the very quick? We all know *economical people* who seem to be niggardly about the amount of air they breathe and the amount of appreciation they will allow themselves to give to anything. They shrivel body and soul. Economy is waste; it is waste of the juices of life, the sap of living."[25]

In 1921 New York retailers launched the National Prosperity Committee to combat thrift. Posters from this period read "Full Speed Ahead! Clear the Track for Prosperity! Buy What You Need Now!" The explicit arguments against thrift that began appearing in popular magazines included strongly worded polemics: "Miserliness is despicable," wrote C. W. Taber in 1922, "hoarding is vulgar; both are selfish, fatal to character and a danger to the community and nation."[26] As the "fun morality" personified by the flapper took hold, the durability or reliability of mass-produced goods like automobiles was taken for granted by a new generation of consumers, most of whom were women, and they now turned their attention to comfort, luxury, and prestige in the products they bought.

In Greenwich Village, a Bohemian lifestyle of gratification replaced old-fashioned restraint, and by the time industrialists and advertisers were finished exploiting this movement, it had spread across much of America. *Exile's Return*, Malcolm Cowley's 1934 memoir of his "literary odyssey in the 1920s," described "Village values" that were at the core of the roaring twenties: "It is stupid to pile up treasures that we can enjoy only in old age, when we have lost the capacity for enjoyment. Better to seize the moment as it comes, to dwell in it intensely, even at the cost of future suffering. Better to live extravagantly."[27]

The movement against thrift was an essential precursor to psychological or fashion-based obsolescence, the second developmental stage of product obsolescence. By the 1920s the habit of conserving worn goods for reuse was challenged on a variety of fronts. Hoarding had become a bad word, as Americans fetishized the new. In the period between the two world wars, stodgy older values, including durability and thrift, were gradually rejected by government officials, engineers, and the general public. In 1913 Postmaster General Frank Hitchcock had pinched pennies by making government pencils last; by 1944 the largest, most expensive disposable product of the century was rolled out—Tiny Tim, a single-use booster rocket. By 1948 Project Hermes had launched its first Bumper-WAC rocket, an event made possible by the Tiny Tim. And in the following year, 1949, the company that had invented the Yankee dollar pocket watch changed its name to Timex and began to manufacture disposable sealed-movement wristwatches. Time was marching on.

Each year the new crop of automobile offerings casts into obsolescence the used and unused models of the previous year . . . The greater visibility of the automobile brings into play the added impetus of rivalry with neighbors and friends.

PAUL MAZUR, *AMERICAN PROSPERITY* (1928)

2   The Annual Model Change

The practice of deliberately encouraging product obsolescence grew out of the competition between Ford and General Motors in the 1920s. The corporate leaders involved in this contest, Henry Ford and Alfred Sloan, both trained as electrical engineers, but there all similarities ended.

Ford began life as a farm boy and became a hands-on engineer obsessed with delivering value and durability to the American public. He first learned his craft in a Detroit machinist's shop, before joining the Edison Illuminating Company. There, he soon became a chief engineer famous for an autocratic style modeled after the "Old Man" himself. Ford saw his car as a great social leveler, a democratic one-size-fits-all symbol of American class-lessness.

Sloan, on the other hand, was the child of privilege—savvy, political, and pragmatic. He graduated at the top of his class from MIT in 1892, a year after Ford joined Edison. For a graduation present, Sloan asked his father to buy him John Wesley Hyatt's bankrupt roller-bearing plant, which he turned into a multimillion-dollar operation. Sloan was less interested in social change than in power and prestige. As head of General Motors, he was ideally placed to achieve both.

MIT gave Sloan the professionalism characteristic of an emerging new generation of engineers. Their values were markedly different from those of Edison and the great inventors of the nineteenth century. In particular, they had a growing awareness of the inevitability of technological obsolescence, a concept that had begun to acquire the ideological trappings of Darwinism. After a series of mergers and takeovers, Sloan found himself at the center of General Motors—a disorganized and crisis-driven company that offered little competition to Ford. At first, Sloan tried to fight Ford with the tools of classic engineering: by making GM cars technologically superior to their competition. Only when this struggle failed did he turn to more creative means of marketing his product.

A CLASH OF VALUES

In the early years of the twenty-first century, when working cell phones and other IT products are discarded by their owners after eighteen months of use, it is difficult to imagine a mass-produced consumer product created without planned obsolescence in mind. But that is exactly the way Henry Ford created his Tin Lizzie.

The Model T was a reliable product marketed at the lowest possible price. For this reason, Ford was able to withstand competition for years. But the durability of the Flivver was problematic to its manufacturer because it postponed repetitive consumption. On average, one of Ford's cars lasted eight years, about two years longer than any other automobile. Unwilling to compromise the quality of his product or modify its external design, Ford faced the challenge of sustaining consumer demand by cleverly manipulating economies of scale. By tailoring his assembly line to a relatively unchanging product design in a single color, he could continually slash prices as unit costs fell in what appeared to be a perpetually expanding market. For years, this strategy worked. Lower costs enabled more and more Americans to purchase a Model T as their first car. But the ceiling on Ford's ability to expand the Flivver's market, and the expense of making even minor modifications to his product, became obvious as automobile sales approached the saturation point.

By 1920, 55 percent of all American families—nearly every family that could afford a car—already owned one. That same year, a minor economic depression resulted in a drastic shortfall in sales for all manufactured goods. This "buyers strike" created a crisis at Ford and at General Motors, which were both in the midst of costly expansions. Ford needed the revenue from Model T sales to pay for its new Rouge River plant. In the coming year, Henry Ford (unlike William Durant at General Motors) successfully resisted borrowing money from a J. P. Morgan consortium to cover his operating and expansion costs. This minor financial miracle left him with absolute autonomy over the Ford Motor Company. But Ford's control was a mixed blessing. He was now more determined than ever to resist changes to his Model T.

Born during the Civil War, Ford held old-fashioned values about engineering, especially the value of product durability. Most engineers in the nineteenth century designed and built their products to last. An incredible example is a hand-blown carbon-filament light bulb, made by Shelby Electric Company, that still illuminates the municipal fire hall in Livermore, California: it was originally switched on in 1901. Although the enormous expense of changing the Model T's design certainly played a part in Ford's resistance to model change, his stubbornness was largely due to his antiquated values. Back on top of the world in 1922, Henry Ford offered the American public an explanation of his refusal to modify the Model T:

> It is considered good manufacturing practice, and not bad ethics, occasionally to change designs so that old models will become obsolete and new ones will have the chance to be bought . . . We have been told . . . that this is clever business, that the object of business ought to be to get people to buy frequently and that it is bad business to try to make anything that will last forever, because when once a man is sold he will not buy again.
>
> Our principle of business is precisely to the contrary. We cannot conceive how to serve the consumer unless we make for him something that, so far as we can provide, will last forever . . . It does not please us to have a buyer's car wear out or become obsolete. We want the man who buys one of our cars never to have to buy another. We never make an improvement that renders any previous model obsolete.[1]

As surprising as this passage is for the vigor with which the father of mass production rejects product obsolescence and force-fed repetitive consumption, Ford's words provide a clue to the source of his popularity and influence in early modern America. He was adamant in his determination to provide American car buyers with more than fair value, and they responded with enormous brand loyalty. The automobile historians Allan Nevins and Frank Hill emphasize Ford's lifelong "preoccupation with durability" and paint a portrait of an idealistic engineer who seems strangely impractical to a modern sensibility: "As a mechanical genius, perhaps the greatest of his time, [Ford] was intensely practical, [but] he had very little interest in competition. The integrity of the product was always the first consideration; consumer demand came second, and any thought of profits was incidental."[2] In his day, Henry Ford stood steadfast against unnecessary obsolescence. He represented an absolute ethic of quality and durability in manufactured goods. Unfortunately, these admirable principles would become the cause of his defeat.

At MIT, Alfred Sloan had learned the opposite lesson: that the dynamism of capitalist economies makes technological obsolescence nearly inevitable. Manufacturers successively improve the technology of their products because these improvements provide them with an edge over their competitors by increasing efficiencies and reducing costs. As a consequence, more effective machines become cheaper to buy. In the brave new world inherited by Sloan and his sophisticated turn-of-the-century classmates, progress toward a technological utopia was accepted almost without question. The rapid succession of inventions that had already improved the quality of life was all the proof they needed. These included telephones, the transatlantic cable, electric street lights, automobiles, and, later, airplanes, radio, and the earliest electric appliances.

The notion of progress, reinforced by followers of Charles Darwin and the "social Darwinist" Herbert Spencer, played a central role in the American consciousness of the day. In academia and in the popular press, explicit analogies between biological evolution and technical design proliferated. They permeated the writings of the Chicago architect Louis Sullivan, who coined the phrase "Form follows function" to encapsulate his conception of building design as evolving organically from the requirements of the physical and cultural environment. Three prominent American scholars, William Ogburn, S. C. Gilfillan, and Abbott Payton Usher, developed theories of technological progress that rested heavily on Darwinian thought.[3] The advertising cliché "New and improved" dates from the earliest years of the new century and captures the idea that products advance in response to changing market competition, much as species evolve in response to changing habitats. With a frequency that was alarming to old-fashioned inventors like Henry Ford, machines, like species, were becoming suddenly extinct.

After Thorstein Veblen published his *Theory of the Leisure Class* in 1899, this technological extinction became popularly known as "obsolescence," a word that Veblen particularly liked to use.[4] Beginning with General Electric, manufacturers invested in research and development departments whose express mission was to produce "the next best thing," and in the process—inevitably—hasten product extinction. Wooed away from universities with financial incentives and promises of freedom to experiment, scientific researchers in these companies sometimes described their inventions as first, second, or third generation, according to the history of their innovation and how recently they had become obsolete. Sloan, who was determined to use technological obsolescence to its best advantage, relied on scientific research in the design of his GM automobiles—until he discovered a more competitive way to do battle with Henry Ford.

When Pierre DuPont took control of General Motors in 1920, it was a mess, following years of crisis management by William Durant. At that time, Chevrolet marketed an inexpensive but unpopular car with a troubled design and flawed engineering. It was no competition at all for the Model T, whose market share stood at 61 percent in 1921. At first GM considered scrapping the whole Chevrolet division. But DuPont, who had graduated from MIT in chemical engineering in 1890, listened to the star graduate of the 1892 class who was now a rising GM executive.

At Sloan's suggestion, DuPont handed over the task of giving GM cars a technological edge to a proven innovator by the name of Charles F. Kettering, the man who had patented the electric starter in 1913. The starter had doubled the potential market for automobiles by opening the door to women as owners and operators. Overnight, hand-cranked cars were obsolete. By 1921 Kettering, who would now head GM's research efforts, had a new idea which, when attached to an inexpensively priced Chevrolet, might render the Tin Lizzie obsolete also. Kettering called his idea the copper-cooled engine. Exactly 759 of these GM cars were manufactured, but after their release early in 1923 the company was swamped with complaints about noise, clutch problems, wear on cylinders, carburetor malfunctions, axle breakdowns, and fan-belt trouble.[5]

The source of most of the problems seems to have been a lack of organizational communication and cooperation at GM. In any case, Sloan became president of GM in the spring of 1923, just as this crisis was brewing. He had already made a fortune by exploiting applications for roller bearings, a novel and radical technology of the time. In an irony few people appreciate, the earliest Model Ts used ball bearings from Alfred Sloan's factories. Henry Ford had been Sloan's best customer for twenty years. All the while Sloan had studied his future competitor.[6]

Sloan liked Kettering, and he loved Kettering's plan to render all Model Ts obsolete. Pragmatically, however, he quickly terminated development of the first air-cooled car following an vigorously negative performance report to GM's executive committee.[7] As a result, no air-cooled engine was reintroduced to the American automobile market until the 1950s. Faced with a technological innovation that would not work, Alfred Sloan gambled that style alone might prove an effective way to compete with Henry Ford for the remainder of the 1923 buying season.

With nothing to lose, Sloan offered his customers a quickly and superficially improved Chevrolet. GM engineers hastily reworked the car's major mechanical flaws; but most importantly, they completely *repackaged* the car, in an era that did not yet have a word for packaging. Almost overnight, the car's lines were made low and round, in imitation of the luxury cars of the day, and its hood was elongated to suggest that it contained a powerful engine. Next to this redesigned version of the 1923 Chevrolet, the Model T looked like a piece of farm machinery. Car customers quickly noticed the difference and responded favorably to GM's prestigious design, as well as its competitive pricing.[8]

Sloan was a quick learner. The '23 Chevy's remarkable success convinced him that mechanical or technological obsolescence was just one of many marketing strategies that he could use to sell new cars. Over the next few years, as he refined his notion of obsolescence, he saw that style could date cars more quickly and reliably than technology. In manufacturing terms, psychological obsolescence was superior to technological obsolescence, because it was considerably cheaper to create and could be produced on demand.

DESIGNING FOR STYLE

The climate of affluence in the 1920s and the contemporary assumptions of the American car buyer made the time exactly right for the introduction of fashion into the manufacture of automobiles. With mechanical quality now more or less a given, people became interested in sophisticated design and presentation, especially those Americans who had been exposed to European culture during World War I. England and America had been forced during the war to challenge Germany's monopoly on dye manufacturing, and as a result cheap color for industrial and textile uses became readily available after the armistice.[9] Almost simultaneously, potential customers, especially women, became frustrated and bored with the clunky, monochromatic Model T.

As early as 1912, automobile manufacturers had begun accommodating the needs and sensibilities of women by making minor changes. In addition to Kettering's electric self-starter in 1913, these changes included upholstery, windshields, interior lights, closed, noiseless (and odorless) mechanical compartments, and separate, closed, roomy (and later heated), passenger compartments. But despite women's demands for comfort and styling, Henry Ford steadfastly refused to prettify his Tin Lizzie. Its reliability—greater than any nonluxury car before or since—and its exponentially declining unit cost made the Model T a popular workhorse in rural America. But Lizzie was also bumpy, noisy, smelly, and homely—traits that more and more women in the 1920s wanted to avoid. Soon the Model T was the butt of jokes, and ridiculed in songs and cartoons.

Ford's ad men did what they could to attract women owners and operators. To provide Ford's closed-body models with a touch of class, copywriters for Ford undertook a project they called "the English job," renaming their coupe and sedan the Tudor and the Fordor, respectively. In the October 1924 issue of the *Delineator*, the headline of a genuinely beautiful ad guaranteed "Freedom for the woman who owns a Ford." The illustration depicted a young woman in pants (an assertion of her modernity) collecting autumn leaves at the roadside. Another ad from the same year shows a young woman being greeted by her mother as she arrives home for the summer holidays—Ford's reliability was especially valued on long trips. Unfortunately, this ad made claims for the Ford that American women of the 1920s knew were outrageous: "The attractive upholstery and all-weather equipment . . . suggest comfort and protection on long trips, while the simple foot-pedal control assures ease of operation in crowded city traffic . . . An increasing number of women . . . who prefer to drive their own cars, are selecting the Ford Fordor . . . as a possession in which they can take pride."[10]

Simple foot pedal control, a growing female clientele, and especially pride in Model T ownership were claims that raised eyebrows among contemporary women. "Humility," wrote Stella Benson, tellingly, "is the first thing expected of a Ford owner." An English travel writer, Benson naively toured the United States in a Model T for her honeymoon in the same year that the Ford ad appeared. After criticizing the car's noise and appearance, she offered a woman's unfavorable impressions of the Tin Lizzie's nearly "suspension-less" comfort: "I had been a wreck owing to . . . [the] constant jolting, which left me so violently giddy . . . I could at no time stand without support, and sometimes could not even sit upright . . . A hotel in El Paso . . . refused a room . . . because I gave the impression . . . I had already called in all the ninety-seven saloons of Juares. Deming, New Mexico . . . found me reeling and rolling still. As for the really beautiful steep rusty city of Bisbee, Arizona, its high vivid mountains whirl and swing upon my memory like great waves of the sea."[11]

Profound discomfort aside, the single feature of the Model T that most troubled contemporary women was its black color, which they considered a cipher of its low prestige. The memoir of an automobile executive (ironically, a Ford employee) confirmed the importance of color to women in automobile selection: "Negotiating with a dealer for a used Lincoln car, [I] limited [my] questions to mechanical subjects. But when [I] got home, [my] wife demanded: 'What kind of upholstering has it?' [I] replied, 'I did not pay much attention to that.' 'Well,' she pursued, 'what is the color of the paint?' . . . [I] responded, 'I think it is a sort of blue . . . ' At this point [my] wife gave . . . a glance of withering scorn. What I was interested in . . . was an automobile that would run. The other party was interested in looks."[12]

For women purchasing a new car, good looks (especially color) were not a new consideration. In 1918 the Jordan Motor Car Company had introduced the first woman's car, the Sports Marine. Advertising copy written by company president Edward S. Jordan in 1917 described the lack of choice in automobile color and the importance of this issue to contemporary woman: "I stood on the corner of Fifth Avenue and 42nd Street in New York, and watched the motorcars . . . Nearly every one of them appeared to be in mourning, finished in dark, repelling shades of black and blue—black hoods and blue bodies. Only a few of the high priced cars appeared in striking shades . . . It is true that while men buy cars, women choose them [with a] quick glance at the body . . . catching the appeal of a striking color."[13]

But color also presented an enormous problem to manufacturers. Varnishes that could suspend enough color to finish an automobile had to be applied by hand and dried very slowly; and they were easily damaged by rain. By 1911 Ford switched to spray-painted enamels and force rooms (drying ovens) that baked the finish onto cars. Unfortunately, this process discolored the pigments suspended in the enamel. In 1914 Ford simply stopped offering the Model T in any color except black. This strategy succeeded because no large competitor could offer a comparatively priced alternative. By the 1920s, Dodge and GM also had a black-only policy for their highest volume models.[14]

In 1921, when control of GM shifted from William Durant to DuPont, all of this changed. DuPont was primarily a chemical and dye-making firm, and in 1918 Pierre DuPont had convinced his board to invest heavily in GM stock, partly because of the guaranteed outlet the car company would provide for their finishes. Late in 1921 a DuPont lab accident revealed that pigment would suspend nicely and dry very well in a nitrocellulose solution. After two more years of experimentation, Duco lacquer was unveiled on the first GM product, the True Blue Overland of 1924. This was the same year that Cutex debuted its first colored nail polish. And like Duco, the rose-colored pigment on the fingernails of stylish American women was a nitrocellulose product.[15]

Despite the Overland's modest popularity, the real success story of 1924 was the new model Chrysler, a fast and relatively inexpensive six-cylinder car that was available in a variety of colors and styles. Alfred Sloan was paying attention. In addition to manufacturing its own car with six cylinders shortly thereafter, GM started to change the color and appearance of their cars on an annual basis. At first, this annual model change was superficial in an engineering sense, but it was significant enough to be noticed by consumers. The new strategy was not simply to make the Model T appear dated. That had already been accomplished. Now, Sloan worked at outdating the styling of GM's *own* earlier models, in order to encourage consumers to trade in their GM cars and buy new ones. Or, more accurately, trade *up:* GM began to offer graduated product lines that encouraged customers to enter a new class of prestige and comfort each time they made a trade. Nothing could have been further from Henry Ford's vision of a classless American transportation device. If Model Ts had been demo-cratic levelers, GM cars were now becoming social stratifiers.

As Sloan's ideas about obsolescence evolved during the waning years of the 1920s, he incorporated other marketing refinements from many different fields, including women's fashions, retail sales, and deficit financing. And as Ford Motor Company's market share dwindled to 30 percent, Henry Ford reluctantly got the point that the rest of the world saw his ubiquitous product as an antique. He realized that if he did not introduce a newer, better-looking car soon, Ford Motor Company would not survive. Under one of Ford's former employees, Big Bill Knudsen, Chevrolet's sales were increasing steadily, as new plants were constructed ac-cording to Knudsen's flexible model of mass production. Unlike the system at Ford's new Rouge River plant, the machinery in Chevrolet plants was not dedicated to a single purpose or design. This flexibility of Chevrolet's assembly line made it easy to accom-modate continual model changes. The output of Chevy plants in-creased from 25,000 automobiles in January of 1926 to 77,000 in November.

The success of GM's flexible mass production strategy must have been especially galling to Henry Ford, since he had rejected Knudsen's suggestions in 1920. Perhaps more troubling than the realization that Knudsen had been right was the knowledge that Sloan had listened to Knudsen and made the savvy decision to hire him, whereas Ford had turned his ideas away. But even now, with the Model T rapidly losing ground, Henry Ford was still un-willing to commit to the major expense and inconvenience of producing an entirely new car for a changing market. Instead, he took small steps to forestall disaster. In 1924 he made superficial changes to the Model T's body, and in 1925 he reintroduced color. Although the toy maker Modine had sold pressed-steel toy versions of Flivver trucks and sedans in its Buddy "L" series for years, the 1925 version of the Tin Lizzie was commemorated in a larger, expensive, pressed-steel Tootsie-toy in a variety of contemporary colors.[16] Unfortunately, the toy's nationwide popularity was not reflected in sales for the Model T itself, which continued to decline.

Tin Lizzie sales were undercut by the fact, now widely known, that Ford engineers were experimenting with newer models. Some of Ford's press releases hostilely denied this, carrying headlines such as "Ford To Fight It Out with His Old Car."[17] The company's joint policy of secrecy and disinformation only excited press enmity, costing Ford valuable credibility and good will. But the Model T's designer and manufacturer was desperate to protect the declining market value of an enormous backlog of cars. In Ford's secret half-formed plan, sales revenue from these automobiles would go directly into a war chest to pay for the expensive retooling to come.

Late in 1926, Ford himself denied publicly that a major model change was imminent. "The Ford car will continue to be made in the same way," he stated flatly. "We have no intention of offering a new car . . . Changes of style from time to time are merely evolution . . . But we do not intend to make a 'six,' an 'eight' or anything else outside of our regular products. We are not contemplating any extraordinary changes in models." This statement, similar to the outrageous claims made in Ford ads targeted at women, was probably not taken very seriously, even by Ford's loyal customers.[18]

THE 1927 CADILLAC LASALLE AND THE MODEL A

Alfred Sloan had been as pleased with Knudsen's success as he was delighted with Ford's continuing difficulties. Through psychological obsolescence, GM's president had guaranteed that his company would remain America's premiere automobile producer for decades to come. Having none of his competitor's scruples about product durability, Sloan did his utmost to find new ways to decrease durability and increase obsolescence. But Sloan was also much less of an autocrat than Henry Ford. His ability to find the right people, like Knudsen, and then to genuinely listen to what they had to say were great strengths in a volatile market.

In 1926, playing to the competitive edge that styling had given GM in the conflict with Ford, Sloan took the advice of the president of his Cadillac division and hired Harley Earl, a custom car designer. Earl made distinctive roadsters for Hollywood's elite, including a car for cowboy star Tom Mix that had a saddle on its roof, and a more subtle $28,000 creation for Fatty Arbuckle that has been described as "softly sculptural." In 1921 Earl built six Cadillac sports sedans for then-Cadillac president Richard Collins. Earl's large frame and flamboyant personality won him notice, and he became a regular guest at West Coast Cadillac parties. One night, Earl bragged to Collins's successor, Larry Fisher, that he could make a Chevrolet look like a Cadillac. It was 1925. Fisher thought for a moment and then offered Earl a job.[19]

Earl's custom work blended all of the visible features of the car's body into one harmonious design. In the industry, this was recognized as a characteristic of the very best luxury cars of the day—the Hispano-Suiza, for example—and it was something Sloan and Fisher wanted to incorporate into their Cadillac line. At Detroit's GM headquarters, Earl revealed the secrets of his technique to GM executives. He developed five full-size models for Cadillac, sculpted entirely out of clay. Other designers of the period worked in wood and metal, but clay gave Earl the flowing forms that made his custom creations unique.

Overall, Earl's models lowered the body and lengthened the

wheelbase. The cramped passenger compartment at the rear of the car created an atmosphere of romantic intimacy, while the length of the car's body (especially when it was disproportionately assigned to the engine space) created an impression of mechanical power and speed. This impression of length and strength was accentuated by the lowered wheelbase, which was the antithesis of practical car design, given the bad roads of the day. Earl was obsessed with the horizontal "through line" in his cars' design. This through line was achieved by integrating the belt line under the side windows into an unbroken horizontal rule that ran the length of the car. Earl used a special device called a highlight gauge to measure the angle at which light reflected from a given car's through line.[20]

The 1927 Cadillac LaSalle had strong luxury lines. Earl had relied heavily on the design of the Hispano-Suiza to create a truly beautiful custom sports car. His LaSalle offered much of the Hispano's styling and most of its features, but at $2,500 it was one sixth of the $15,000 price tag. The success of Harley's LaSalle in the 1927–28 market confirmed Sloan's strategy of emphasizing styling and fashion change in automobile marketing. In a little known article in *Printer's Ink*, Sloan himself wrote: "More attractive products are coming into the market continually and influence the purchaser to exchange his car a year or more old for a new car of the latest design."[21] Delighted with the LaSalle's success, Sloan soon created the first styling department at an American automobile manufacturer. It was called the Art and Color Section, and Sloan appointed Harley Earl to be its head. Thereafter, Earl busied himself in creating incremental modifications in GM cars for the annual model change.

Because it was too expensive for GM to change each model completely every year, major redesigns requiring new dyes were put on the three-year styling cycle that would eventually define the lifespan of all so-called durable goods in America. Between these major styling changes, annual face-lifts rearranged minor features, such as the chrome work. But even these minor moves created the illusion of progress and hastened the appearance of datedness that psychological obsolescence required. Years later, Earl openly discussed his role in creating what he called planned or dynamic obsolescence: "Our big job is to hasten obsolescence. In 1934 the average car ownership span was 5 years: now [1955] it is 2 years. When it is 1 year, we will have a perfect score."[22]

In the spring of 1927, after the fifteen-millionth car rolled off the line, Ford Motor Company shut down production of the Model T forever, in order to retool for the Model A. Since Alfred Sloan was not yet a household name, the *New York Times* described what was happening as "the fight for the national automobile championship between Henry Ford and General Motors."[23] When Ford's new car was revealed on November 30 of that year, the American press swooped down on Dearborn, Michigan, to watch Charles Lindbergh, the nation's most recent hero, demonstrate the Model A's modern features. Publicity photographs of the event depict a youthful, clear-eyed Lindbergh sitting tall behind the wheel of the new Ford Tudor with an elegant older woman smiling graciously beside him. She is Gertrude Ederle, Queen of Romania.

With the joint endorsement of American and European royalty, the Model A became an overnight success. Its features included safety glass in the windshield, as well as aluminum alloy pistons, heat-treated chromium steel gears, and anti-friction bearings throughout—all of which made the car run quieter and smoother. The body was cushioned against the chassis at all points with rubber and hydraulic shock absorbers to make the ride comfortable for the ladies. The nine and a half inch clearance to the road gave the car fashionably low lines, and all seven body types were available in four colors. The Fordor sedan was available in seven. To top it all off, the price of the Model A was lower than that of comparable Chevrolet models. Four hundred thousand orders for the new car poured in nationwide before any dealership saw delivery of the new cars.[24]

Historians do not know for sure how much the Model A cost Ford Motor Company. Estimates of the design costs alone range as high as $18 million. Added to that was the cost of completely shutting down production for six months in order to retool and produce the new car. Ford later guessed that his total costs were in the region of $100 million. But some historians regard this low figure with skepticism, putting the real cost of the Model T's obsolescence, including shutdown, loss of sales, and complete retooling, at about $250 million.[25]

Whatever the actual number was, the Model A was a very expensive lesson in psychological obsolescence, but one that Henry Ford still did not completely grasp. Despite the fact that Ford hired a new agency to create ads promoting the Model A as a "smart and stylish car" and depicting it in elegant settings in order to create a classier image, America's foremost advertising historian reminds us that Ford was still unwilling to accept the full implications of psychological obsolescence and repetitive consumption: "On the eve of its unveiling, he contradicted the implications of his advertising by proclaiming his intention to make the new car 'so strong and so well-made that no one ought ever to have to buy a second one.'"[26]

Gradually, as the novelty of the Model A wore off, Ford sales declined once again. In 1930 cosmetic changes were introduced to the Model A body types. But still, Harley Earl's increasingly stylish

Chevrolets, with their six-cylinder engines after 1928, wore down Ford's market lead. In *Auto Opium,* a comprehensive and readable study of the history of automotive design, David Gartman observes that by 1931 Ford sales were one third of their 1929 level, forcing the company once again to shut down production and revamp its line.[27]

What came next must have made Alfred Sloan smile. In 1932 Ford introduced the first low-priced V-8 in fourteen different models. The Model A had proven to be a costly and unnecessary interim step in Ford's reeducation. Ford production now went over to the GM strategy of creating superficially different models on standardized running gear. Still, Ford sales fell again in 1932. The following year the company finally adopted GMs policy of changing the style of its cars regularly on an annual basis. Psychological obsolescence was now the rule for U.S. automakers. And because car production was America's flagship industry, this lesson was quickly copied in all other areas of manufacturing.

PRODUCT ADDICTION

The creation of the Art and Color Section at GM was one of a flurry of design events in 1927 that indicated profound changes in U.S. industry. That same year, Egmont Arens left his position as managing editor of *Creative Arts Magazine* to devote himself full time to designing a line of Art Deco lamps. His choice was not as strange or trivial as it might at first seem. In August, Arens's future employer, Earnest Elmo Calkins, president of Calkins and Holden Advertising, published an influential article in *Atlantic Monthly,* to which he was a regular contributor. "Beauty: The New Business Tool" popularized the growing trend toward professional product design.

Historians usually trace this trend to the influence in America of the 1925 Paris Exhibition (officially named the Exhibition Internationale des Arts Décoratifs), and especially to the beautiful glass pavilion exhibits by René Lalique. But the influence of the Paris Art Deco show was actually part of a larger design awareness in America following World War I. In 1924, a full year before the Paris show, the *Annals of the American Academy of Political and Social Science* devoted a special issue to "Scientific Distribution and Modern Selling." This widely read collection included a piece by Huger Elliot, head of Philadelphia's School of Industrial Art.

In what was really a manifesto calling for professionalism in industrial design, Elliot described how America's city planners, architects, textile workers, furniture designers, and silversmiths could "contribute to the daily increase of beauty" and raise the visual standards of Americans. Clearly, Elliot had a democratic vision: "These producers create for the man who cannot afford to buy 'museum pieces'—the objects which will enable him to create beauty in his home—the furniture and silver with good, simple lines, the rugs and hangings fine in color. How important is the mission of these designers. How necessary that they be well trained! How imperative that they realize the great part they play in forming the taste of a nation!"[28]

Calkins's article appeared three years later, in 1927. It was a contemporary observation of the transition then taking place in American design sensibilities. "Back in the mauve decade, or the gay nineties," Calkins wrote, "when a manufacturer produced a machine . . . it never occurred to him to . . . make his device pleasant to look at as well as efficient." Calkins blamed "the persistent influence" of the Puritans in creating suspicions about beauty and encouraging the widespread belief that ugliness guaranteed technological integrity. America has changed, Calkins noted: "We enjoyed our era of the triumph of the machine, we acquired wealth, and with wealth education, travel, sophistication, a sense of beauty; and then we began to miss something in our cheap but ugly products. Efficiency was not enough. The machine did not satisfy the soul . . . And thus it came about that beauty, or what one conceived as beauty, became a factor in the production and marketing of goods."[29]

Calkins took specific aim at Henry Ford as a symbol of the old-fashioned, anti-aesthetic style of manufacturing:

> In those days . . . Henry Ford began making his famous car. It was an honest piece of work, a motor car that functioned at an unbelievably low cost, thought it did violence to three senses, sight, hearing and smell; but people in those days were unable to forget long enough their wonder that the thing should be to mind the intrusion of more ugliness into a world that was losing peace and silence and the beauty that inheres in old things. And so the Ford car was put out, and chugged along faithfully on all our roads. The public laughed at it and christened it "Lizzie," but bought it and used it in increasing numbers, and Mr. Ford rested secure in his belief that he had solved one of the major problems of human existence and that there was nothing more to be done.

Months before Sloan created GM's Art and Color Section and right after Ford Motor Company had closed its Model T assembly line to retool for the Model A, Calkins provided his analysis of how Ford lost the battle to psychological obsolescence. Style had become the newest and most important selling feature of the day: "What has happened, apparently, is that many more people have become conscious of style and the style idea has been extended to many more articles . . . People buy a new car, not because the old one is worn out, but because it is no longer modern. It does not satisfy their pride . . . You cannot make people substitute a new car that runs well for an old car that runs well unless it has some added quality. The new quality must be borrowed from the realms of good taste, smarter lines, newer design, better color, more luxurious upholstery, more art, or at least more taste."

Calkins was an advertising executive with an uncanny marketing sense. Despite his shift to the word "taste" in the final sentences of this passage, the keyword for obsolescence of style in this passage is really "pride." Manipulationist theories of consumption are not as popular among sociologists today as they once were, but few would deny that psychological obsolescence was a strategy designed to put the consumer into a state of anxiety based on the belief that whatever is old is undesirable, dysfunctional, and embarrassing, compared with what is new.

Obsolescence of style—a specialized kind of psychological obsolescence—focuses consumer attention on the visual or design features of conspicuously consumed personal items, ranging from cars, cell phones, clothing, hats, jewelry, laptops, lighters, and luggage to PDAs, pens, pocket knives, purses, shoes, sunglasses, and watches. In a consumer culture, people size one another up continually to establish status hierarchies based on disposable income and taste. If a person has money to purchase the latest items of self-presentation, he or she seems superficially more affluent and therefore presumably more socially successful, more desirable. From a salesman's perspective, such people are also the best prospective customers. Because they willingly bought previous models, they are much more likely to purchase newer and newer consumer items.

The other side of this pride and self-presentation equation is shame, or more precisely the anxiety about feeling shamed that creates a state of watchfulness in American consumers for what-ever is new. The basic idea in shame-based advertising is that the desire not to lose face can be manipulated to produce conspicuous consumption. This idea is as old as the sumptuary laws that became the basis for the emergent seventeenth-century fashion industry.

Thorstein Veblen first formulated what would become known as conspicuous consumption in 1899. It is important to remember, however, that Veblen's formula concerned the "vicarious" consumption of the leisure class. In Veblen's model, aristocratic men created wealth, which their wives consumed and displayed. In a society based on vicarious consumption, the wealth producer is distanced from the shame the consumer (his wife) experiences whenever she appears unfashionable in society.[30] But by 1920 Americans were confronted with such an abundance of goods that conspicuous consumption could not remain vicarious. This abundance came about in part because electricity had begun to replace steam as the driving force of industry. During the next quarter of a century, capital productivity would increase 75 percent, while labor productivity would grow at an even faster rate.[31]

The habit of conspicuous consumption in order to either feed one's pride or reduce one's shame is frequently referred to as "product addiction." One of the few industrial designers to write openly about America's product addiction was Frank Lloyd Wright's most famous apprentice, Victor J. Papanek. Tragically, in the United States product addiction begins early, Papanek observed: "Latent product junkies first get hooked when they are still babies. Toddlers playing with shoddily made, badly designed toys learn that things exist to be thrown away and replaced by anything 'new' . . . There is nothing wrong with children playing with dollhouses, small cars or baseball cards. What is disturbing is the hard-core advertising that uses these toys to encourage children to own, collect, and ultimately risk becoming product addicts."

Papanek went on to say that while American products once set international standards for quality, consumers of other nations now avoid them due to shoddy American workmanship, quick obsolescence, and poor value. Scant resources and increasing pollution were making the American practice of consumer waste a destructive strategy that was costly to pursue. Papanek also saw immense human costs to force-fed repetitive consumption that were much more difficult to quantify: "Millions of people have substituted the satisfaction of owning things and spending money for any meaningful reward in life. Most things are not designed for the needs of people, but for the needs of manufacturers to sell to people."[32]

Papanek was writing in 1983. Today, such explicit statements about the costs of product addiction are almost as rare as insights into its mechanics. Analyses of how advertisers and designers collude to manipulate consumers into buying new goods are now very difficult to find, even in modern advertising textbooks. This was not the case in the 1920s, when the basic techniques of "manipulationism" were being developed. Back issues of *Printer's Ink* and *Advertising and Selling* abound with practical advice from which many contemporary copywriters and graphic artists benefited. They often included caveats, like the following warning about being too heavy-handed: "Such subjects are delicate ones to attack in an advertisement. If there is a single carelessly chosen word, there is apt to be resentment at the intrusion, the covert insinuation."[33]

This same article from 1928 recounts how silver manufacturers mounted a cooperative advertising campaign to "shame the prospect into buying the latest model of a venerably old product."

American newlyweds' habit of prizing their heirloom silver was preventing repetitive consumption, so "it was obviously necessary for us to make the people who cling to the old sets realize just how out of date they are. Ridicule of the past from which the silver was handed down proved to be the best plan. Any manufacturer of a quality product will tell you that the article that refuses to wear out is a tragedy of business." An example of how shame was used to market wristwatches can be found in the Elgin series of magazine advertisements: "The objective of this entire campaign was to cause owners of old-style watches to be self-conscious concerning them and to go out and buy the latest type of watch. The copy was by no means afraid to suggest discarding an ancient Elgin, by the way."

This self-conscious concern about being out-of-fashion is the key feature of psychological obsolescence. Anything that is unfashionably dated is psychologically obsolete, but psychological dating can depend on features other than style or design. Businesses that sell an experience, such as watching a movie, rather than an item to be taken home and used have had to push product dating beyond the limits of style-based marketing. This fact became obvious when America's entertainment industries looked for strategies to encourage psychological obsolescence.

AT THE MOVIES

Like the annual model change, the Academy Awards did not begin as a marketing strategy, but they very quickly became a rating system to encourage repetitive consumption. As movie production became increasingly technical and expensive, producers eagerly introduced innovations that rendered previous kinds of films obsolete. This was obvious in 1927, with the introduction of sound recording as well as crane and tracking shots. As production budgets for films like *Sunrise* and *The Jazz Singer* began to expand, ticket prices increased, and the movie industry made an effort to move up-market. The first Academy Awards, held in May of 1929, honored the films of the pivotal 1927–28 season. That first year featured only twelve categories of awards, and *Wings* (with a record-setting production budget of $2 million) became the only silent film ever to win best picture (called Best Production in those days).

After *Wings*, sound rendered silent films technologically obsolete. But it was psychological obsolescence rather than technological innovation that would ultimately drive the Academy Awards. The publicity surrounding the Oscars involved moviegoers in a public competition among films. The nomination process, and the period of delay between the announcement of the nominees and the award ceremony, encouraged Americans to see all movies nominated in a particular category before the award deadline in order to judge for themselves which film should win. Moreover, the award itself gave new life to Oscar-winning films. As the perennial Oscar hopeful Martin Scorsese put it, "When people see the label Academy Award Winner, they go to see that film."[34]

Other entertainment industries gradually took a lesson in psychological obsolescence from the clever strategy of the film industry. *Billboard*, a screen and show business magazine, was first. At the height of the Depression in 1932, it developed an early version of a weekly hit parade for prerecorded radio songs and sheet music. By the time its song-chart format was formalized in 1944, *Billboard*'s listing had become a major record marketing tool and a positive force for subscription and advertising sales.

Billboard's success did not go unnoticed in the rest of the publishing world. During the height of the summer book season in

1942, the *New York Times* established a similar device that encouraged psychological obsolescence on a weekly basis in the book industry. The first list, called "The Best Selling Books Here and Elsewhere," was more subtle than the annual automobile model change in the way it deployed datedness to market books and newspapers.[35] But, like a hit parade song chart, it depended on psychological obsolescence. Every Sunday for over fifty years, the *New York Times* bestseller listing has encouraged the repetitive consumption of books by surveying what is in and out of fashion. This list has made the *Times* (and its imitators) attractive both to consumers at the subscription end, who want to know what they should be reading, and to book publishers at the advertising end, who allocate huge budgets to trumpet their newest releases and pump up sales.

We are living in the midst of that vast dissolution of ancient habits.

WALTER LIPPMANN, *A PREFACE TO MORALS* (1929)

3   Hard Times

Justus George Frederick was an advertising man's advertising man. Somehow, as a Pennsylvania farm boy from a large family, he developed a love for the books that had been rare items in his rural German-speaking home. From childhood it was obvious that Frederick had a gift for clear, enthusiastic writing. He started professional life as a printer's devil in Chicago, but soon became editor of the budding advertising trade magazine *Printer's Ink*. As a skillful and ambitious copywriter, he attracted the notice of the J. Walter Thompson ad agency, which recruited him and sent him on to their New York office in 1908. In Manhattan, Frederick edited another advertising trade journal, *Advertising and Selling*, before he left to establish his own business press, The Business Bourse.

Frederick wrote novels, cookbooks, management and economic manuals, as well as reams of advertising copy and advertising news. He helped found the Advertising Men's League of New York and co-founded the League of Advertising Women with his wife, Christine Frederick. Of the two, Christine was much more famous and successful in her own day, but J. George, as he was known, is now generally recognized as the man who invented progressive obsolescence.[1] This puts him at the head of a group of writers who by the early 1930s were devoting considerable attention to products that were made to break—a group that would include the distinguished company of Archibald MacLeish, Aldous Huxley, and Lewis Mumford.

Justus George first introduced the concept of progressive obsolescence in a lead article for *Advertising and Selling* in the fall of 1928. With characteristic energy, he wrote: "We must induce people . . . to buy a greater variety of goods on the same principle that they now buy automobiles, radios and clothes, namely: buying goods not to wear out, but to trade in or discard after a short time . . . the progressive obsolescence principle . . . means buying for up-to-dateness, efficiency, and style, buying for . . . the sense of modernness rather than simply for the last ounce of use."[2]

Progressive obsolescence was J. George's attempt to reshape America's thinking about the social role of advertising and design following the obsolescence of the Model T and the introduction of the Model A. Instead of coining a new word, Frederick stuck with the familiar "obsolescence," which had achieved notoriety in 1927, when "even Ford [had] been forced to bow before the god of obsolescence."[3] But he smoothed down its negative connotations by combining the word with the most positive term available in that remarkable age of social and technological innovation, progress itself. In advocating "progressive obsolescence," Frederick was trying to elevate Sloan's practice of annual model changes to an economic habit that would sustain America's economy by means of perpetual repetitive consumption and growth in all industries.

At a time when consumerism was still an unfamiliar term,

Frederick had a firm grasp of the concept. He encouraged manufacturers to target American shoppers with a cooperative advertising campaign that would alter the nation's buying habits: "The consumers to be reached . . . are the twenty millions . . . who can now afford to buy on the progressive obsolescence principle and who already do so in some lines. They should be faced with the powerful logic and attractiveness of practicing more rapid obsolescence in their purchasing."[4]

J. George's campaign was riding on the coattails of two other men, Joseph Alois Schumpeter and Paul Mazur. In 1928 Frederick had come into contact with the ideas of an Austrian-born economics professor at the University of Bonn through Mazur, a Manhattan business acquaintance. Schumpeter was refining a view of capitalism he had set forth in 1912 in his *Theory of Economic Development* (a work that would not appear in English translation until 1934).[5] Late in 1929, in the wake of Black Tuesday, Schumpeter's notion of business cycles would provide a shocked and desperate world with insight into economic devastation, and his "creative destruction" would describe forms of obsolescence that might fuel capitalism's recovery.[6]

Although Frederick, in his 1928 article, did not credit Schumpeter's work, J. George's notion of progressive obsolescence was very similar to the forces of perpetual market change that drove capitalism in Schumpeter's model: "The fundamental impulse that sets and keeps the capitalist engine in motion," wrote Schumpeter, "comes from the new consumers' goods . . . that capitalist enterprise creates . . . The same process of industrial mutation—if I may use that biological term— . . . incessantly revolutionizes the economic structure from within, incessantly destroying the old one, incessantly creating a new one. This process of Creative Destruction is the essential fact about Capitalism."[7]

At least one other German-speaking New Yorker understood Schumpeter's theories sufficiently to write about them. In January 1928, Paul M. Mazur addressed the Advertising Club of New York on the topic of psychological obsolescence in order to publicize his forthcoming book, *American Prosperity: Its Causes and Consequences.* That book, published in March, was immediately reviewed in *Printer's Ink.* Mazur—an investment banker and partner at Lehman Brothers—was well-rounded in economic theory, and well-connected in New York, even writing occasional essays for the *New York Times.* J. George, the ambitious farm boy from Pennsylvania, envied Mazur's cosmopolitan accomplishments and listened thoughtfully to his ideas. He would later "borrow" much of what Mazur wrote, including the comparison of obsolescence to a god (the same god to whom Ford had already bowed). In Mazur's original phrasing, "Wear alone . . . [is] too slow for the needs of American industry. And so the high-priests of business elected a new god to take its place along with—or even before—the other household gods. Obsolescence was made supreme."[8]

But Frederick borrowed much more than rhetorical flourishes from Mazur's book. One sentence in particular would preoccupy J. George and Christine for the next two years. "If what had filled the consumer market yesterday could only be made obsolete today," Mazur wrote wistfully, "that whole market would be again available tomorrow."[9] It was exactly this problem of stepping up the pace of repetitive consumption that J. George sought to address with his notion of progressive obsolescence. "American genius" now makes it possible, Frederick wrote a few months after Mazur's book debuted, "to possess a marvelous cornucopia of interesting products of which there is such a great sufficiency that it may not only be owned in one model to last a lifetime, but in a kaleidoscopically rapid succession of colors, tastes, designs and improvements."[10]

This was convincing copy, and it had an impact in the New York business world of 1928. In the year leading up to the Wall Street crash, five out of twenty issues of the American Management Association's journal were devoted to the use of art, color, design, or fashion in business strategy.[11] Companies began to establish design departments or to contract the services of industrial designers. In 1929 Raymond Loewy got his first commission redesigning the genuinely ugly Gestetner duplicator, and that same year Walter Dorwin Teague created his famous Art Deco Brownie for Kodak. But of more immediate concern to J. George was the publication in 1929 of the book that would transform his wife, Christine, into the Martha Stewart of her day.

As business books go, *Selling Mrs. Consumer* was a runaway success. It rode a wave of popular concern about what goods and styles women (the purchasing agents of their families) most wanted. It also claimed to tell businessmen how to sell to women, since, as Christine claimed, there is a "very real difference between men and women in purchasing habits and consumption of goods."

> Woman are far heavier consumers of personal goods than men, utilizing the principle of obsolescence far more frequently and naturally. Second, [there is a] greater love of change in women. We are only beginning to see that there is tremendous significance in all this; and that America's triumphs and rapidity of progress are based on progressive obsolescence. We . . . have an attitude that is quite different from the rest of the world, and . . . we have been speeding it up . . . It is the ambition of almost every

American to practice progressive obsolescence as a ladder by which to climb to greater human satisfactions through the purchase of more of the fascinating and thrilling range of goods and services being offered today.[12]

Christine's willingness to collaborate with industry in targeting female consumers earned her some hostility from the first wave of feminist historians, some of whom considered her a "double agent."[13] But recent opinion has reached a more balanced perspective. Like Henry Ford, Christine, with her provincial background, solid education, and considerable intelligence and talent, was caught between the nineteenth-century ideals and the emerging twentieth-century realities in the male-dominated business world. Although she hated housework, Christine challenged herself to create an at-home career in consumer and domestic research so that she could stay close to her children as they grew up. In 1935 she was named among the thirty most successful women in Greater New York.[14]

Similar conflicts between old and new values surfaced again and again in personalities of the 1920s. So much of the world was then in transition, as new things constantly replaced old ones, and so many old values were coming into conflict with new ones— perhaps this helps explain why "obsolescence" became such an expressive and powerful concept during the late 1920s and 1930s. In their everyday lives, ordinary people were becoming familiar with the need to discard not just consumer goods but ideas and habits that had suddenly became obsolete. In her explanation of progressive obsolescence, Christine gave insight into this new mentality. It involves, she wrote, three telltale habits of mind, all very amenable to fashion change:

(1) A state of mind which is highly suggestible and open; eager and willing to take hold of anything new either in the shape of a new invention or new designs or styles or ways of living.

(2) A readiness to 'scrap' or lay aside an article before its natural life of usefulness is completed, in order to make way for the newer and better thing.

(3) A willingness to apply a very large share of one's income, even if it pinches savings, to the acquisition of the new goods or services or way of living.[15]

Possibly because they were blinded by the rapid changes of the day, few people foresaw the devastation to the world economy that would occur late in October 1929. Like so many others, J. George Frederick, a man professionally sensitive to social trends and fashions, missed it completely. He spent 1929 touring Europe with Christine, publicizing her new book and writing his own *Philosophy of Production*, a two-volume work. The subtitle of Book Two, *Whither Production?* would turn out to be a bitter pun in the anxious economy of 1930, where production had indeed withered. With their money, possessions, and jobs gone or in jeopardy, few Americans wanted to hear from J. George that "there has been an imperfect realization of the sound and genuine philosophy in free spending and wasting . . . many people still are a little shame faced about it. They drink today the remnants of yesterday's milk, rather than today's fresh milk, fearing that they may be wasting. They cut themselves off from stimulations and pleasures on an obviously false plea that they 'can't afford it.' They worry along with old equipment when improved or new equipment would actually be an economy, or would add to their leisure and fullness of experience."[16]

Book One, which included essays by Christine, Earnest Elmo

Calkins, and even Henry Ford, was a dismal failure and became another source of strain in the Fredericks' troubled marriage. J. George began to spend more time in the expensive New York offices of The Business Bourse, just off Times Square. Christine—now the family's primary breadwinner, through royalties and speaking engagements—lived with the children at Applecroft, their Green Lawn, Long Island, home, which she transformed into a domestic research laboratory. By 1930 J. George and Christine were comfortably on the road to complete estrangement.

WINDOW SHOPPING

Although the Fredericks' phrase "progressive obsolescence" passed into total obscurity in the first full year of the Depression, the practice itself was, ironically, still very much alive.[17] As competition for consumers' few dollars intensified in all fields, manufacturers were eager to use whatever means they could to encourage people to buy their product rather than someone else's. Design competition became the standard American business strategy of the day, and style obsolescence began to dominate corporate thinking about products as diverse as radios, cameras, furniture, kitchen appliances, men's shoes, plumbing fixtures, silverware, fountain pens, cigarette lighters, and compact cosmetic cases. With the Depression, the direction of American industry passed from the hands of engineers into the hands of designers. Design frenzy soon extended beyond renewable products to include elevator interiors, trash barrels, locomotives, and skyscrapers, of which the best known example is New York's Chrysler Building of 1930, still beautifully festooned with its Art Deco hubcaps.

Sigfried Giedion observed that the power of industrial designers "grew with the Depression" until all mass-produced objects bore the stamp of designers whose "influence on the shaping of public taste [was] comparable only to that of the cinema."[18] America's discovery of the visual was vast. What is now recognized as the golden age of comic books came right after the crash. Science fiction, detective stories, and jungle adventures developed into full-length graphic novels during this period. A new aesthetically constructed round of advertising appeared in magazines, including (my personal favorite) the Cadillac series in *Fortune* magazine during the fall of 1930. The new visual sensibility was expressed in the art and photography of the day, which produced some of the finest works of social realism and abstract expressionism, and in the quest to develop television. During the height of the Depression, the pioneer inventors Philo Farnsworth and Vladimir Zworykin became locked in a patent battle over the iconoscope, a prototypical television camera. Photojournalism came to the United States from Europe with the publication of *Life* magazine in 1936. *Life* itself provided a popular venue for the industrial photography of innovators like Margaret Bourke-White.

In short, every aspect of America's visual culture, high and low, expanded during the Depression. In the years immediately following the crash, as money became tighter and America became more visually literate, its citizens took vicarious pleasure in a new national pastime, window shopping—a phrase coined during this period. A new profession also emerged: consumer engineers, who were a combination of what we now know as industrial designers and product placement or marketing specialists. The manifesto of this new profession came out of Calkins & Holden, the premier advertising agency for visual design in America at that time. In *Consumer Engineering: A New Technique for Prosperity* (1932), co-authors Roy Sheldon and his boss, Egmont Arens (the new head of the Calkins & Holden design department), relied heavily on the ideas of J. George Frederick, an old friend of the firm's president, Earnest Elmo Calkins. In their manifesto, Sheldon and Arens wrote that "Consumer Engineers must see to it that we use up the kind of goods we now merely use. Would any change in the goods or the habits of people speed up their consumption? Can they be displaced by newer models? Can artificial obsolescence be created? Consumer engineering does not end until we consume all we can make."[19]

Hesitating a little over phrasing, Sheldon and Arens did not use the term "progressive obsolescence," which Christine Frederick herself had once described as "pompous."[20] Instead, they wrote about the difficulty of providing obsolescence with an appropriate name. Previous names like progressive waste, creative waste, or fashion carried negative connotations because they focused on the scrapping or junking process, which made consumers and manufacturers nervous.[21] In the desperate economic conditions of 1932, the modern manufacturer would have to realize that obsolescence "is more than a danger . . . it is also an opportunity . . . [at first] he saw obsolescence only as a creeping death to his business. But now he is beginning to understand that it also has a possible value . . . it opens up as many new doors as . . . it closes . . . for every superseded article there must be a new one which is eagerly accepted."[22]

In order to circumvent the negative connotations of obsolescence, Sheldon and Arens followed J. George's lead in coining a new term that would itself be extremely short-lived. In their defense it should be said that they were primarily designers by profession, not copywriters: "Obsoletism," they write, is a "device for stimulating consumption. This element of style is a consideration in buying many things. Clothes go out of style and are replaced long before they are worn out. That principle extends to other products—motorcars, bathrooms, radios, foods, refrigerators, furniture. People are persuaded to abandon the old and buy the new in order to be up-to-date, to have the right and correct thing . . . Wearing things out does not produce prosperity, but buying things does."[23]

OBSOLETE MAN

With so many Americans suffering joblessness and deprivation during the Depression, a band of critics began to decry the growing trend toward the mechanized replacement of manpower. The early focus of this critique was that most visible symbol of mechanization, the coin-operated vending machine. The explicit purpose of these machines, which proliferated after 1928, was to replace human sales clerks. In 1932 *Billboard* magazine recorded that such machines give "the appearance of taking away jobs from people who might work."[24]

That same year, *Fortune* published an anonymous essay condemning the "technological unemployment" that had led to "a serious decline in the number of wage earners in basic industries." This essay marked the first time that "obsolescence" was used to describe the social reality that human workers could be replaced by machines. "Obsolete Men"—like Jonathan Swift's *A Modest Proposal*—contained bitter satire: "For some two or three millions of years the world's work was done by a patent, automatic, self-cooling mechanism of levers, joints and complicated controls with a maximum life of about three score years and ten, an average efficient working day of eight to twelve hours, an intermittent power production of one-tenth of one horsepower, and certain vernal vagaries for which there was no adequate explanation in the laws of physics . . . [In] its honorable function as a producer of primary motive power, it is now not only outmoded but actually obsolete."[25]

The essay was written by Archibald MacLeish, who was then at the height of his powers as a poet and on the verge of winning the first of three Pulitzer prizes, this one for *Conquistador,* a poem cycle about the Spanish conquest of Mexico. From 1920 to 1938 MacLeish served on *Fortune* magazine's editorial board before becoming Librarian of Congress in 1939. He was a keen observer of American culture and could write hauntingly about American society:

> It is . . . true that we here are Americans:
> That we use the machines: that a sight of the god is unusual:
> That more people have more thoughts: that there are
>
> Progress and science and tractors and revolutions and
> Marx and the wars more antiseptic and murderous
> And music in every home . . .[26]

MacLeish was a member of the League for Independent Political Action, an organization that sought alternatives to industrial capitalism, which league members saw as an unnecessarily cruel social system. His *Fortune* essay on human obsolescence resonated with a sense of the betrayal that Americans felt about the abrupt end to the first period of abundance in their history and the ensuing years of deprivation: "The decade of mechanical marvels ended in the depression of 1929. We had been informed that the mechanization of industry and the resultant increase of production led necessarily to lasting plenty. And when we stumbled over the bluff of November, 1929, we could hardly believe our senses. We blamed government. We blame the expansive manufacturers. We blamed our own speculations. We blamed—we are still blaming—the bankers. But it never occurred to us to ask whether the blame might not more properly be attached to fundamental changes in industrialism itself."[27]

MacLeish's article established that while the creation of new industries had in the past absorbed workers who had been displaced by machines, a turning point had come and steps must now be taken to prevent the "apparent trend toward rapidly increasing unemployment in the future." The two steps that appeared possible were a decrease in productivity by "a kind of legislative sabotage," or an increase in the consumption of goods "by a change in the technique of distribution."[28] It was this second vague solution that MacLeish favored.

"Obsolete Men" was clearly influenced by the economic analysis of the contemporaneous "technocracy" movement. And like the criticisms of the technocrats, MacLeish's acute criticisms of the industrial system were unaccompanied by specific recommendations about what to do (though he did commend "the share-the-work movement for rationing the residuum of employment among the employed and the unemployed by the introduction of the five-day week").[29] In the fall of 1932 when MacLeish's *Fortune* article appeared, the technocracy movement's short-lived popularity was at its peak.

Although its earliest origins were in a group called the Technical Alliance led by Thorstein Veblen's protégé, Howard Scott, technocracy really began in May of 1932 when an ad hoc group, the Industrial Experimenters Association, met briefly at the invitation of Walter Rautenstrauch, a professor of engineering at Columbia University.[30] Rautenstrauch was convinced that the ultimate cause of the Depression was the profit motive—the inability of businessmen to curb their quest for profit in the interest of social harmony. And since businessmen could not be counted on to exercise self-control, Rautenstrauch believed that engineers must take up the responsibility of reorganizing supply and demand. In Rautenstrauch's view, the social machine was broken, and broken machines were amenable to an engineering fix.[31]

As he would do with any engineering problem, Rautenstrauch tackled the Depression by conducting an exhaustive survey of industry in order to find an empirical solution to its breakdown. At about this time Rautenstrauch's friend Bassett Jones introduced him to Howard Scott, a Greenwich Village personality and trained engineer who had been sporadically conducting just such a survey for ten years. Both men blamed capitalism's "price system" for the ills of the Depression and felt that only engineers, with their grounding in scientific methodology, were intellectually equipped to steer American industry toward a course of prosperity.

By the fall of 1932 the United States was grasping at an increasing number of straws in its desperate search for an end to the Depression. As a fourth year of economic blight loomed, Americans' primary topic of conversation was "What are we going to do?" But after the first interviews with Scott appeared in the *New York Times* in late August, the national topic of conversation shifted. The new question on many people's minds was "What is technocracy, and could that be the solution to our problems?" The technocracy movement seemed much better than many of the other utopian schemes of social reform proliferating at the time, though the topic was poorly understood. Still, it offered hope to desperate people, and for a brief spell the country went "technocrazy."[32]

The basic ideas of the movement were that new technologies had thrown capitalist America into a depression and that the price system had outlived its usefulness, because it empowered businessmen and politicians whose interest in maximizing profit was incompatible with the promise of a technological society. The crisis of the Depression required that society be restructured by engineers and economists around the principle of production for the use and prosperity of the many, rather than the profit of the few.[33]

Most of the utopian plans—technocratic or otherwise—that emerged during that troubled year of 1932 spoke of the need for a body of experts who would restructure society so as to achieve a balance between supply and demand. Such a balance would eliminate technological unemployment. But to the ears of America's business community, what technocrats advocated in a variety of pamphlets like the Continental Committee's *Plan of Plenty* began to sound genuinely threatening. Herbert Hoover's defeat in the November 1932 elections and Roosevelt's loud promise of something called "a New Deal" already had them feeling vulnerable. Following the publication of MacLeish's article in December 1932, with its prediction of manpower obsolescence, a shadowy effort to discredit Howard Scott and the technocrats took shape.

It began with a series of attacks on Scott by Allen Raymond in the *New York Herald Tribune*. Scott proved an easy target, since he had made fantastically inflated claims about his background and training. By early January 1933, Charles Kettering, Alfred Sloan, and an assortment of America's business leaders openly criticized technocracy in the press. A convincing lead article in the *New York Times* on January 8, along with a second article in the *New York Post* on January 13, deflated most of the major claims of the movement. An embattled Scott was urged by Rautenstrauch and the others to fight back by addressing a banquet of industrialists, economists, and artists at New York's Hotel Pierre. The most extensive nationwide radio hookup created to that date provided Scott with a forum to respond to criticisms. Unfortunately, he had no training as a public speaker, and rancorous dissension among the technocrats themselves, compounded by a long month of personal attacks, had taken their toll on his spirit. Although he was usually a very commanding presence, Scott came across terribly on the radio.

Rautenstrauch and the other major players broke publicly with Scott ten days later, and he was prevented from continuing his industrial survey at Columbia University.[34] Under Scott's faltering leadership, technocracy struggled on as a Depression-era movement, but after January 1933 its momentum was lost, and it became a fringe movement.

THE BUSINESSMAN'S UTOPIA

Like technocracy, "planned obsolescence" was conceived during the desperate year of 1932. And in its early incarnation, it too focused on restructuring society around a body of experts whose mandate was to achieve an equilibrium of supply and demand that would eliminate technological unemployment. Unlike technocracy, however, planned obsolescence was not a movement. It was the idea of one man, a successful Manhattan real estate broker by the name of Bernard London.

London lived well south of Columbia University, on elegant Central Park West, but his extensive contacts in Manhattan's architectural, Jewish, Masonic, and academic communities may have included early members of the technocracy group.[35] All we really know about him is that he began his career as a builder in Russia. His love of architecture and the pleasures of reading enabled him to pursue a vigorous lifetime project of self-education in the history of building construction. Naturally enough, this led him to the real estate business, as it had, some thirty years earlier, led him to membership in the Masons. The Masons (or Freemasons) were among the world's oldest and largest fraternal organizations, dedicated to "making good men better" through rituals and instruction involving the symbols of architectural craftsmanship. Perhaps it was through his fellow Masons that some of the ideas of the technocrats became available to London at quite an early date, well before Scott gave his first interview to the New York newspapers on August 21.[36]

During the summer of 1932, a sharp decline in real estate sales left London—fifty-nine years old and at a point of transition in his life—with plenty of time on his hands. As he sat in his offices on East 40th Street near Times Square with little to do, London wrote a twenty-page pamphlet called *Ending the Depression through Planned Obsolescence*. The business examples that pepper this document came almost exclusively from real estate, but London also drew on his wide reading in making his case.[37] It is not clear how he distributed the pamphlet or to whom, whether he gave it away free or charged money in those tight-fisted times, and indeed whether London invented the phrase "planned obsolescence" or whether it was already circulating in New York's business community. What we do know is that London used the phrase in the title of his first publication in 1932, giving it—to whatever limited extent—exposure during the Depression. Over twenty years later, the Milwaukee designer Brooks Stevens would claim to have invented planned obsolescence himself (see Chapter 6), but Stevens's claim does not stand up to scrutiny.

In his first pamphlet, London outlined a scheme that combined features of technocracy with the kinds of commercial obsolescence that were familiar from the work of Sheldon, Arens, and the Fredericks. "The essential economic problem," as London saw it, was "one of organizing buyers rather than . . . stimulating producers." London was dismayed that "changing habits of consumption [had] destroyed property values and opportunities for employment [leaving] the welfare of society . . . to pure chance and accident." From the perspective of an acute and successful businessman, the Depression was a new kind of enforced thrift:

> People generally, in a frightened and hysterical mood, are using everything that they own longer than was their custom before the depression. In the earlier period of prosperity, the American people did not wait until the last possible bit of use had been extracted from every commodity. They replaced old articles with new for reasons of fashion and up-to-datedness. They gave up old homes and old automobiles long before they were worn out, merely because they were obsolete . . . Perhaps, prior to the panic, people were too extravagant; if so, they have now gone to the other extreme and have become retrenchment-mad. People everywhere are today disobeying the law of obsolescence. They are using their old cars, their old tires, their old radios and their old clothing much longer than statisticians had expected.[38]

In order to combat the social ill of the "continued planless, haphazard, fickle attitudes of owners," London recommended that America "not only plan what we shall do, but also apply management and planning to undoing the obsolete jobs of the past." London wanted the government to "assign a lease of life to shoes and homes and machines, to all products of manufacture . . . when they are first created." After the allotted time expired,

> these things would be legally "dead" and would be controlled by the duly appointed governmental agency and destroyed if there is widespread unemployment. New products would constantly be pouring forth from the factories and marketplaces, to take the

place of the obsolete, and the wheels of industry would be kept
going . . . people would turn in their used and obsolete goods to
certain governmental agencies . . . The individual surrendering
. . . a set of old dining room furniture, would receive from the
Comptroller . . . a receipt indicating the nature of the goods
turned in, the date, and the possible value of the furniture . . . Re-
ceipts so issued would be partially equivalent to money in the
purchase of new goods.[39]

Like the technocrats, London felt that the government should
empower boards of "competent engineers, economists and math-
ematicians, specialists in their fields," to determine "the original
span of life of a commodity." If only we would accurately add "the
elements of life and time to our measurement of what we produce, and say that the life of an automobile shall be not more than
5 years, or the life of this building shall not last more than 25
years, then, with the addition of our customary measurement of
these commodities, we will have a really complete description of
them right from the beginning."[40]

A scheme that, at first glance, seems today like a crackpot version of progressive obsolescence mixed with a fair measure of
technocracy begins to make a kind of workable sense when reread
in the context of 1930s economic desperation. Bernard London
was obviously an educated man with a good understanding of
business issues, and to his credit he was much less interested in
disenfranchising capitalists and empowering engineers than were
the technocrats. London was primarily interested in achieving an
equitable and workable arrangement between capital and labor:
"When capital purchases the automobile or the building, it will be
doing so only for that limited period of years, after which the re-
maining value left in the product will revert to labor, which pro-

duced it in the first place, and which thus will receive its rightful share in the end, even if it did not do so in the beginning."[41]

London's *Ending the Depression through Planned Obsolescence* was written in the same year that *Brave New World* was published, and London's description of product obsolescence closely resembled some aspects of Aldous Huxley's work. For example, Huxley wrote of the year 600 AF (after Ford: about 2463 AD), "Every man, women and child [is] compelled to consume so much a year in the interests of industry" and then to discard it so that new goods can be manufactured and consumed. Hypnopaedia or sleep teaching indoctrinates the young utopians in the values of a society based on obsolescence by repeating over and over 'Ending is better than mending . . . old clothes are beastly. We always throw away old clothes. Ending is better than mending, ending is better . . . Ending is better than mending. The more stitches, the less riches; the more stitches . . . "[42] Huxley's Utopia was run by Controllers (London used the word Comptroller), and there were a few other minor textual parallels. But if London used Huxley as a model, he may have done it unconsciously, since London's short essay was devoid of *Brave New World*'s acid attack on consumer society:

> Why is it prohibited? asked the Savage . . .
> The controller shrugged his shoulders. "Because it's old; that's the chief reason. We haven't any use for old things here."
> "Even when they beautiful?"
> "Particularly when they're beautiful. Beauty's attractive, and we don't want people to be attracted by old things. We want them to like new ones."[43]

Business historians usually assign the phrase "planned obsolescence" to the 1950s.[44] Whatever his eccentricity, London predates this assumption by twenty years, although it is true that, for London, planned obsolescence did not have the modern meaning of achieving a death date by manipulating the physical structure and materials of a product. To him, a product's death date was exclusively a limit imposed externally by a committee of experts and then enforced as a social rule.

DEATH DATING

Internally imposed or structural death dating was not a product of the Depression. Like branding, it first appeared much earlier, in the nineteenth century. But, as Bernard London noted, people were especially tight-fisted during the hard times of the 1930s, and they tried to eke out the last bit of use from things they had wantonly discarded in previous years of abundance. In such difficult market conditions, manufacturers began to systematize and apply scientific research methods to the loose group of manufacturing tricks they had simply called "adulteration" in an earlier time.

To "adulterate" originally meant to dilute a product in the simplest of processes, in the way that a whiskey trader might add water to his booze. This lowered costs and allowed a lower price, which increased sales. But adulteration was considered a shabby business practice around the turn of the century, and punitive measures were taken to reduce or eliminate it.[45] In 1921 the economist J. A. Hobson described "the struggle of the State to stamp out or to regulate the trades which supply injurious or adulterated foods, drinks, and drugs." Eventually, "adulteration" was used to describe the production of shoddy manufactured goods whose inferior materials and workmanship not only lowered costs but increased repetitive consumption as the product broke or wore out quickly: "A manufacturer or merchant who can palm off a cheaper substitute for some common necessary of life, or some well-established convenience, has a . . . temptation to do so. For . . . the magnitude and reliability of the demand make the falsification unusually profitable."[46]

In *The Tragedy of Waste* (1925), Stuart Chase described adulteration no longer as dilution but exclusively as a habit of greedy manufacturers who wasted labor "through its employment upon materials that have the shortest possible life; upon cloth that goes the soonest into tatters, upon leather that tears and cracks, upon timber that is not well seasoned, upon roads that fall into immediate decay, upon motors that must be junked in a few years, upon houses that are jerry built, upon nearly every article manufactured in quantity for the American public." Like Hobson, Chase connected American manufacturers' practice of adulteration with their attempts to encourage repetitive consumption and with the consequences for an industry that did not practice adulteration: "In the case of the tire industry, quality and wearing power have been increased [circa 1924] to an average life per tire of 1 year and 8 months as against 1 year and 4 months in 1920. The Cleveland Trust Company, in its official bulletin of September 15, 1924, remarked: 'These figures explain some of the troubles that have beset the tire industry, which has been penalized for the marked success of having improved its product.' How penalized? By slowing turnover, and loss of sales."[47]

The Depression gave manufacturers a new incentive to systematize their strategies of adulteration and to apply scientific research methods to the practice of death dating or planned obsolescence in order to encourage repetitive consumption. At first these practices had no name, and even in internal corporate documents manufacturers were reluctant to refer to their own product policies as adulteration, because in order to adulterate a product successfully a manufacturer had to have effective control over his market—either through a monopoly or through a cartel powerful enough to fix prices and standards. If he did not have control over the market, of course, any other manufacturer could compete vigorously by producing a better, similarly priced, product. But monopolies and cartels were illegal by the 1930s, owing to a series of antitrust laws that were increasingly enforced. Moreover, the meaning of "planned obsolescence" had not yet crossed over from its external technocratic use to become an internal industrial substitute for adulteration; and the alternative phrase, "death dating," would not be invented until 1953 (see Chapter 6).

In 1934 Lewis Mumford described practices that would later be called "death dating," but he did not use that term nor the phrase "planned obsolescence." So, despite Bernard London's pamphlets, planned obsolescence probably did not achieve currency among industrial designers until after the 1936 publication of an article on "product durability" in *Printers' Ink*.[48] Still, Mumford recognized death dating and psychological obsolescence for what they were, and wrote books criticizing both practices. Twenty years before Vance Packard would shout about it from the rooftops in the *Hidden Persuaders* (1957), Mumford had this to say: "No one is better off for having furniture that goes to pieces in a few years or, failing that happy means of creating a fresh demand 'goes out of style.' No one is better dressed for having clothes so shabbily woven that they are worn out at the end of the season. On the contrary, such rapid consumption is a tax on production; and it tends to wipe out the gains the machine makes in that department."[49]

Because the practice of planned obsolescence had monopolistic ramifications, the earliest scientific tests to deliberately limit product life spans have left few traces in the public record. General

Electric seems to have been the first American company to devote significant resources to industrial research and development under the direction of MIT Professor Willis R. Whitney. Whitney's mandate at GE was specifically to develop patentable technologies that would allow GE to maintain a monopolistic control of the American electronics market by rendering other manufacturers' products obsolete. He was influential in training the generation of engineers who instituted obsolescence as the standard corporate practice. These included Irenee DuPont, Alfred Sloan, and Paul Litchfield of Goodyear Rubber.[50]

Obsolescence had become a highly personal subject for Whitney in 1904 when his most important invention to that date, the GEM lamp, was threatened by the tantulum lamp developed in Berlin by chemist Werner von Bolton at the Siemens and Halske laboratories.[51] Whitney responded by setting his research team the task of designing a new lamp that would have significant advantages over both his own GEM invention and Von Bolton's tantalum line. By 1907 William D. Coolidge had designed the tungsten filament, which was the basis of our modern light bulb. At the end of a bitter patent dispute in 1920, GE emerged with monopolistic control of light bulb manufacturing for generations to come.

At some point after the beginning of the Depression, GE realized that its monopolistic control of lamp production permitted the company to adulterate the lifespan of its bulbs. During a later antitrust case against the company (U.S. v. G.E., Civil Action 1364), a memo surfaced indicating that GE labs were experimenting with controlling product life spans in order to increase consumer demand and repetitive consumption. In fact, they had been doing so since the crash: "Two or three years ago we proposed a reduction in the life of flashlight lamps from the old basis on which one lamp was supposed to outlast three batteries, to a point where the life of the lamp and the life of the battery under service conditions would be approximately equal. Some time ago, the battery manufacturers went part way with us on this, and accepted lamps of two battery lives instead of three. This has worked out very satisfactorily."[52]

The memo went on to say that such experimentation was, in fact, continuing. It recommended a deliberate shortening of the product life span of flashlight bulbs or one-life battery lamps, since, "if this were done, we estimate that it would result in increasing our flashlight business approximately 60 percent. We can see no logical reason either from our standpoint or that of the battery manufacturer why such a change should not be made at this time."[53]

This memo, dated 1932, described the practice of planned obsolescence without naming it, though the policy it encouraged was not implemented at that time. Still, all the elements of what would become known as planned obsolescence or death dating were clearly in place by 1932. These include the connection between adulteration and repetitive consumption, the term "planned obsolescence" itself, and scientific research into various structural means for limiting a product's life span. All that was needed was for them to come together. When exactly this happened is unclear, but by 1950 this combination had long since taken place.

> If this new philosophy of obsolescence means protecting the people,
> then we have to go and learn our economics all over again . . . Many of
> the things that happened in this country have happened as a result of
> obsolescence.
>
> DAVID SARNOFF, SENATE HEARING TESTIMONY
> (APRIL 11, 1940)

4   Radio, Radio

From the perspective of a culture whose most serious challenge may well be the toxicity of its voluminous waste, a time when the nation's most urgent technological question was how to reduce radio static is hard to imagine. However, the remarkable innovations that eventually solved this problem brought deliberate product obsolescence into the world of radio manufacturing and broadcasting. And from there, planned obsolescence would spread to an entire industry of consumer electronics.

In the early 1930s David Sarnoff, the unstoppable head of RCA who had made a fortune in AM radio, expected to make radio-listening itself obsolete in one vast sweep, by introducing television to the American public and then cornering the market. Television would replace radio just as radio had (he thought) replaced the phonograph, and in both cases RCA would be the winner. To Sarnoff's dismay, obsolescence came to radio technology through two sets of inventions that were maddeningly beyond his control.

One set of inventions eliminated static and permitted realistic, high-fidelity radio broadcasting for the first time in history. The second set of inventions replaced vacuum tubes and wire circuitry with transistors, paving the way for space-age miniaturization and a consumer electronics industry that developed independently of the giant radio manufacturers. The young customers who purchased these convenient portable devices accepted them as inherently disposable. And the savvy businessmen who made them lost no time in planning for their continual obsolescence and replacement.

FM BROADCASTING

In 1906, on one of many visits to New York, Guglielmo Marconi befriended an enthusiastic office boy who was doing odd jobs in the Front Street lab of technician Jimmy Round. Perhaps Marconi, who had experimented with electromagnetism in his parents' vegetable garden as a teenager, recognized something familiar in the kid from Hell's Kitchen who was treated as an equal both by the American technicians and by the 'Coni men aboard the Italian inventor's ships. David Sarnoff got many things from his friendship with Marconi, including dedication to broadcasting, advanced training in electrical engineering, administrative responsibilities, and probably even his tendency to womanize. Marconi's friendship also left Sarnoff with the unfailing ability to recognize and befriend genius wherever he found it. He recognized it in a fellow Russian émigré, Vladimir Zworykin, RCA's television pioneer. He recognized it too in a New York native, Edwin Howard Armstrong, the father of FM radio, with whom he developed a powerful and tragic rivalry.

Not a man to back down from life's strongest experiences, "Major" Armstrong loved five things with total—some would say foolhardy—abandon: radio, his wife, his country, fast cars, and the thrill of free climbing in high places. Patents on regenerative circuits he developed in 1913 as a gifted Columbia undergraduate brought him considerable income from RCA and recognition in the field. The next year, as a postgraduate fellow, Armstrong began a thorough investigation of radio static with his mentor, Michael Idvorsky Pupin, in Columbia's Marcellus Hartley Research Lab.[1] Both men agreed that static interference was the biggest problem plaguing AM broadcasting, but eight years later, after they had exhausted every possible avenue of attack, they admitted they were stumped.

In the meantime, though, Armstrong had became a multimillionaire by selling two additional patents to David Sarnoff at RCA, and he had fallen in love with Sarnoff's beautiful darkhaired secretary, Marion MacInnis. One day Armstrong offered her a ride in the fastest car he could find. When she agreed, he imported a fawn-colored Hispano-Suiza from Paris, then took her for a trip around Manhattan and through the boroughs. After a whirlwind courtship, Armstrong married Marion in 1923 and drove her in the Hispano-Suiza to Palm Beach for their honeymoon.[2]

Although he was lucky in love, Armstrong was less lucky in the lab. During an acrimonious patent dispute over regenerative circuits that Armstrong eventually lost in 1928, AT&T had begun to refer publicly to Lee De Forest as the father of regenerative circuitry, and Armstrong felt that his reputation as a radioman was becoming tarnished. Motivated by this setback, he hired two assistants (Thomas Styles and John Shaughnessy) and resumed work on static, applying himself to the problem with staggering determination. From 1928 until 1933, the three men designed, built, and tested circuits for transmitters and receivers relentlessly, seven days a week, from nine o'clock in the morning to well into the night. Armstrong became so absorbed he would forget to change his clothing from week to week. His rumpled ties refused to lie flat on his soiled shirts. He was inattentive to Marion, who, from all appearances, suffered gracefully by cultivating her women friends and sometimes taking long train trips. During this period, at work and at rest, Armstrong could talk of nothing but circuits and measurements.

Gradually, the three men had to admit that an entirely new system of transmission and reception was required. But by 1931 they had not made much headway with their invention. Desperate for a breakthrough, Armstrong began to experiment with widening the frequency of a broadcast signal, something he and Pupin had ignored, since everyone agreed it would be unproductive. Immediately, the critical signal-to-noise ratio improved. Armstrong and his team then concentrated on achieving a 100-to-1 ratio (three times better than AM). Eventually, they developed a three-step process that started with Armstrong's own super-heterodyne circuit. This first circuit was attached to something Armstrong called a "limiter," a new kind of circuit that removed a major cause of static (signal variation). Finally, a second new invention (named a discriminator) returned the limited FM wave to its original electronic form and passed it on to the speaker, where it was reproduced as sound.

This new reception system eliminated atmospheric static completely and also faithfully reproduced a much wider range of sounds than AM. Suddenly, high fidelity radio (or hi-fi, as it would become known by a postwar generation of ex-army techs) became a real possibility. In some of Armstrong's earliest demonstrations, the sound of paper being crumpled, water being poured into a glass, and an oriental gong being struck were correctly identified by audiences seventeen miles away from the microphone that had picked them up. FM was also more energy-efficient than AM. The transmitter for Armstrong's first experimental FM station, W2AG in Yonkers, used the same amount of power as a large electric light bulb. Still smarting from his 1928 lesson in intellectual property rights, Armstrong filed four patent applications on December 26, 1933. And around the same time, he invited his RCA friend, David Sarnoff, soon to be his nemesis, to view his astonishing achievement.

In 1933 Armstrong was still involved in the legal battle with AT&T over rights to his regenerative circuit. RCA stood to lose Armstrong's patents, and indeed the company would eventually take legal action against him to recover some of their losses. But this in no way affected Armstrong's relationship with Sarnoff. As a younger man, Armstrong had often dropped by Sarnoff's home for coffee and the inevitable talk about the future of radio. Sarnoff's children called Armstrong "the coffee man." Later, he and Sarnoff corresponded, and their letters, although formal, express deep mutual respect.

They had become friends back in 1913, when Armstrong first demonstrated regenerative circuitry and arc station signal reception to a small group from Marconi Wireless that included Sarnoff, who was then Marconi's chief assistant engineer. Years later the RCA mogul wrote about his astonishment at being able to hear signals from Honolulu in Pupin's Columbia lab over Armstrong's "little magic box." An element of envy may have entered Sarnoff's relationship with this brave, brilliant, and wellborn New York native, growing stronger, perhaps, after Armstrong married Sarnoff's secretary. But both men still shared a hero in Guglielmo Marconi. In 1929 Armstrong tracked down Marconi's first wireless shack in America. When he found it, near Babylon, Long Island, it was being used as a paint shed. Armstrong bought it and presented it to Sarnoff, who installed it in RCA's biggest station at Rocky Point.

If Sarnoff was envious, he was also grateful, and in gratitude he could be as magnanimous as he was powerful. In 1922 Armstrong's super-regenerative circuits enabled RCA to compete forcefully again in radio manufacturing. Leadership in this area had slipped out of the company's grasp with the introduction of L. Alan Hazeltine's neutrodyne circuit. When RCA bought exclusive rights to Armstrong's super-heterodyne circuit in 1922, they were acutely aware that maintaining their monopoly meant preventing another neutrodyne disaster. Consequently, in drawing up a contract with Armstrong, Sarnoff asked for first refusal on future inventions, and Armstrong happily agreed. Sarnoff sealed the deal with $200,000 cash and 60,000 shares of RCA stock. When Armstrong designed an additional circuit that adapted the super-heterodyne for practical radio manufacture, Sarnoff paid with another 20,000 shares. Thanks to his friend Sarnoff, by 1923 Armstrong was a multimillionaire and the largest single shareholder in RCA.

Sarnoff knew Armstrong was capable of even greater things. In particular, he was well aware that Pupin and Armstrong now knew more about the problem of radio static than anyone else alive. Repeatedly, Sarnoff told Armstrong that what he desired was "a little black box" like Armstrong's regenerative "magic box" that would completely eliminate AM static and give RCA a palpable edge in radio manufacturing for years to come. For the five years that Armstrong worked day and night on static, he believed that what he was about to invent was already committed to RCA. Sarnoff apparently felt the same way.

Armstrong was keenly aware that the company his inventions had revitalized was now a de facto monopoly "with the power to stifle competition in the manufacture and sale of receiving sets," according to the Federal Trade Commission's 1923 report.[3] But Armstrong also knew that the power and resources of this monopoly made RCA an ideal venue for a project like FM, which brought expensive R&D, manufacturing, and broadcasting challenges. So when Armstrong offered Sarnoff a private viewing of his new invention, the RCA mogul drove himself uptown to Armstrong's lab in the basement of Columbia's Philosophy Hall. He had a clear sense that radio history was about to be made, and of course he was right.

Despite Sarnoff's abiding sympathy for real genius, what Armstrong showed him that Christmas season in 1933 was wholly unexpected and more than a little disturbing. FM has since been called a truly disruptive invention, one that posed a massive threat to the status quo of AM radio and to the RCA/NBC cartel in particular.[4] At that time, the NBC network included about one thousand AM stations in which RCA had a 50 percent ownership. Of the fifty-six million working radios in the continental United States, not one could receive FM broadcasting. Sarnoff, whose genius for administration and realpolitik may have exceeded Armstrong's genius in electronics, controlled a financial empire comparable to Microsoft or CNN today. It had enormous executive, judicial, and financial influence. As the mogul's authorized biography later described the FM threat, "If adopted across the board, FM would have canceled out every existing radio receiving set and broadcasting station . . . Obsolescence on such a gigantic scale— not for a new service but for an improvement in the existing service—was carrying a valid principle to an unprincipled extreme."[5]

Although Armstrong clearly understood the revolutionary power of FM to restructure the business of radio, he did not really appreciate the magnitude of the threat FM represented to entrenched industries. He was obsessed with advancing the "radio art" and with nurturing his personal reputation as an inventor; the business of politics or the politics of business were outside his ken. "After 10 years in obscurity, my star is rising again," he wrote with disarming naiveté in July 1934.[6] Armstrong did not comprehend the raw obstructive power that Sarnoff wielded through RCA, NBC, and the FCC, a government institution firmly ensconced in the silk lining of the corporate pocket. Armstrong was slow to see that his friend, a shtetl Jew from Minsk who had adapted to the harsh environment of Hell's Kitchen, was also a rapacious survivor for whom personal power meant more than technological progress or friendship.

Perhaps Armstrong's increasing influence at RCA was at the heart of the rivalry developing between them. Sarnoff had been RCA's *force majeur* since 1919. Under his opportunistic leadership, RCA had evolved from being a corporate cipher that merely held the pooled radio patents of GE and AT&T, to being *the* major manufacturer of radio receivers in America.[7] FM's success would put Armstrong into a proprietary position both as RCA's largest shareholder and as the owner of the company's most fundamental patents. Sarnoff's own role would shrink to that of a mere manager. To a man of Sarnoff's intellect, ambitions, talents, and—to be truthful—disfiguring vanity, this outcome was not acceptable.

Another downside of FM, from Sarnoff's perspective, may have been the competition it would offer to early television. The oft-repeated claim by RCA and the networks that people would not pay for a new kind of radio capable of broadcasting free and high fidelity music might turn out to be false. If it did, FM offered the prospect of greater commercial growth in its early years than television did, since TV would require massive investments by manufacturers, programmers, and customers, whereas FM sets were cheap and FM programming could be adapted from existing AM formats. The danger Sarnoff perceived may have been that FM would present Americans with a more affordable, and more familiar, electronic consumer product at a time when television was still finding its legs.

By 1933 Sarnoff had already decided that television, which was then being developed in RCA labs, would soon make radio completely obsolete—the two could not coexist on the airwaves, he thought. The cessation of radio broadcasting would compel American consumers to toss out their radios and purchase one of RCA's television receivers, which would be considerably more expensive than radios and therefore much more profitable for RCA. Sarnoff called this his "supplantive theory" of business.[8] In reality, it was just a new deployment of technological obsolescence. FM had no place in this scheme. From Sarnoff's point of view, Armstrong had built a better dinosaur shortly before their extinction.

STATIC AT THE FCC

A skilled politician, Sarnoff did not oppose Armstrong openly. Between 1933 and 1935 RCA engineers engaged in a noncommittal period of studying and writing reports on FM. Some of these studies misrepresented the new invention, taking issue with slight or unfounded technical difficulties, but for the most part they were overwhelmingly favorable. Then in 1935 RCA's research and engineering vice president, Dr. W. R. G. Baker, the most sympathetic proponent of FM inside the company, was suddenly and inexplicably let go. RCA soon asked Armstrong to withdraw his equipment from their premium location atop the Empire State Building, to make way for Zworykin's television experiments. In

1936, when the FCC's chief engineer, Charles Jolliffe, made a first annual report to Congress on the progress of the communications industries, he strangely made no mention at all of FM. Following the report, RCA hired Jolliffe away from the FCC to head their frequency allocations committee, whose charge was to get the best possible frequencies from the FCC for television. Thereafter, Jolliffe became a dedicated company man and was intimately involved in Sarnoff's campaign against Armstrong and FM.

Although Armstrong still had no word from Sarnoff on RCA's commitment to his new invention, gradually he was getting the picture. When Harry Sadenwater, a friend at RCA, filled him in on Sarnoff's supplantive theory and RCA's intention to make radio itself obsolete, Armstrong started hiring new staff to help him make FM commercially viable without RCA's help. In January 1936 he told the *New York Times,* "The New Year will undoubtedly witness the installation of frequency modulation transmitters . . . The sole difficulty which remains to be overcome . . . [is] vested interests."[9]

Getting a permit for an experimental FM station proved difficult, however. Armstrong suspected that, like the Jolliffe incident, this was an example of collusion and obstruction on the part of RCA and the FCC, but he could not prove it. His application in January 1936 was turned down. After he hired a lawyer to represent him in what was really a trivial and routine request, the FCC finally granted Armstrong experimental privileges in July.

Alexander Pope once wrote, "Beware the fury of a patient man." By 1936 Armstrong was surely enraged. He found a wooded lot on the Palisades in Alpine, New Jersey, and sold some of his precious RCA stock to buy it. He then designed a 425-foot FM transmitting tower which Sarnoff could see from his office on the 53rd floor of the RCA building directly across the Hudson. Arm-

strong had resorted to similarly personal goads before. During his patent fight with De Forest, for example, he had positioned a flag that bore his regenerative circuit's patent number (1113149) right where De Forest could see it from the windows of his house. Whatever the personal origins of the rivalry between Armstrong and Sarnoff, the conflict was out in the open by 1936 and would grow to enormous proportions over the next eighteen years.

In 1937 Armstrong contracted GE's electronics division to build twenty-five FM receivers for demonstration purposes. The man in charge of this new division was unusually committed to FM. He was Dr. Baker, who Sarnoff had fired in 1935. When preliminary broadcast tests began at Alpine in 1937, General Electric asked Armstrong for a license to manufacture FM radio receivers. These GE radios went into production by December; the first ones off the assembly line, identified today by number, are still highly prized by antique radio collectors. With Baker's encouragement, respect for Armstrong's accomplishment was soaring high at GE. But the company had its own reasons for wanting to help him out. The final decision in a 1932 antitrust suit against GE had prevented the company from manufacturing radios for five years.[10] GE executives regarded the antitrust action as a maneuver orchestrated by Sarnoff through congressional overseers of the FCC. By developing FM alongside Armstrong in 1937, GE would be able once again to compete with RCA in radio manufacture, this time from a new position of strength.

By 1939 Armstrong had his Alpine station on the air with full power. Station W2XMN cost more than a quarter of a million dollars to build, but the clarity, quality, and range of the sound it transmitted were convincing proof that FM radio was several orders of magnitude above anything anyone had ever heard over a wireless receiver before. In the same year, two more private FM

stations began transmission before GE itself commenced broadcast experiments from a site near Albany. The proximity of the Albany and Alpine transmitters confirmed one of FM's most valuable and attractive aspects. On FM bands, overlapping signals do not produce interference. Instead, FM receivers home in on whichever signal is stronger in a given area. This capture feature makes FM very attractive to commercial interests, since it means that stations can be positioned close to one another and share the same frequency.

When GE published these findings, RCA finally began to take notice. They quietly applied for and received FCC permission to build their own experimental FM station. By the fall of 1939, 150 more applications had been filed at the FCC—for many more stations than could be accommodated by the five experimental channels. But the FCC would not allocate more bandwidth for FM.

At the same time that RCA applied to the FCC for an FM station, it asked the commission to make permanent the ten channels that had been provisionally granted to television for experimentation in the 44 to 108 MHz range. These were just above the experimental FM band (41 to 44 MHz) and would have boxed FM in to a restrictive range, making expansion difficult. But Congress had recently appointed a new chairman of the FCC, to amend past problems of influence. A committed New Dealer, James Lawrence Fly had no allegiance to RCA or to the networks. Consequently, Armstrong's luck began to change.

He immediately formed an independent FM broadcasting association and appealed to Fly's FCC for more channel space. Through Baker at General Electric, Armstrong was able to acquire the reports on FM's potential that RCA and Jolliffe had suppressed in 1936. This new evidence seriously harmed RCA's po-

sition, as did the fact that they were already advertising, manufacturing, and selling television sets to receive signals within a spectrum of experimental channels that was still under debate. Fly understood that RCA was trying to pressure the FCC by creating popular demand among new TV owners for the experimental channels. He also understood that all of these owners were, not coincidentally, within range of RCA's tower atop the Empire State Building. Incensed by evidence of his commission's earlier collusion with Jolliffe and RCA to suppress FM, Fly ordered all television off the air until FM's case could be heard.[11]

Although it was a ruling that would not stand for long, the FCC reached the initial decision to award TV's experimental 44 to 108 MHz range to FM. Fly's ruling provided independent FM broadcasters with enough room to create many more new stations if they were needed. The FCC further recommended that, in the future, television sound should rely on FM, the superior broadcasting medium. For a moment, it seemed as though Armstrong had won. "Within five years," he predicted, "the existing broadcast system will be largely superceded."[12]

Sarnoff apparently agreed. In 1940, through an intermediary, he suggested a partnership between Armstrong and RCA to achieve joint monopolistic control of radio manufacturing and broadcasting in general. RCA had a strict policy against paying royalties, but the intermediary, Gano Dunn, an old friend of Pupin's who was now a director at RCA, suggested Armstrong could merge his FM patents with RCA's. RCA would then administer the licenses and fees and split all royalties evenly between them. At the time, Sarnoff was about to appear before a Senate hearing arising out of the FCC findings against RCA.[13] He was obviously in a conciliatory mood, offering to deal Armstrong into television if Armstrong would deal him into FM. But Armstrong demurred, pointing out that such a monopolistic arrangement was likely illegal.

He probably also feared giving RCA administrative control of his patents. Dunn warned him that he could not fight "anyone so powerful as the R.C.A.," but this threat was the company's final negotiating ploy, and all it accomplished was to get Armstrong's back up.[14] Later that same year, when Sarnoff wrote directly to Armstrong's attorneys offering to buy a nonexclusionary FM patent license for $1 million with no further royalties owing, Armstrong rebuffed him. It was a very generous offer, but Armstrong had already licensed Freed Radio, GE, Stromberg-Carlson, Western Electric, and Zenith to manufacture FM receivers in agreements that paid royalties of 2 percent. He offered RCA a more favorable royalty rate of 1.75 percent but would not budge from that position.

Armstrong's intransigence may have been due to the FCC ruling that now required RCA to use FM in all television sound transmitters and receivers. He probably thought that RCA now *had* to deal with him on his own terms. It is possible, however, that he deliberately wanted to prevent Sarnoff from being able to manufacture FM radios and transmitters, just as GE had been prevented from manufacturing AM radios following the antitrust suit of 1932. His personal motive may have been to punish Sarnoff and RCA. If, on the other hand, he was acting strategically, he may have wanted to reduce their considerable size and power. This was certainly the opinion of RCA's vice president of patents, Otto Schairer. Schairer's lawyer advised him that Armstrong's attempt to license the FM system was an "unlawful use of the patent to suppress competition."[15]

By the late 1930s, the contest between Armstrong and Sarnoff seemed to be shaping up as a classic confrontation between a huge corporate entity and an upstart entrepreneur. If Armstrong had conceived of his situation in these terms, he might have acted differently. Advanced capitalism had regularly stifled or swallowed up the individual entrepreneur during these years when the political will to enforce legislation against monopolistic practices was the exception rather than the rule. In 1942 the Harvard economist Joseph Schumpeter would devote a portion of his influential book *Capitalism, Socialism and Democracy* to documenting the causes for what he called "the obsolescence of the entrepreneurial function" in advanced capitalism.[16] But whatever Armstrong's motives were, his decision to exclude RCA proved to be a critical miscalculation, since it left Sarnoff with a strong disincentive to end his campaign against FM.

The forty FM stations and half a million FM receivers in operation in 1942 were still only a drop in the bucket compared to AM, which had over a thousand stations and as many as fifty-six million working receivers. But FM's popularity was beginning to snowball. Even more significant than FM's growth was the fact that RCA began to lose market share in the mobile radio field of special communications systems, such as those used by taxis, fire engines, police, and emergency vehicles. FM's clarity, capture effect, and energy efficiency made it much more suitable to mobile radio than AM. RCA, which had dominated this business until the 1940s, now lost hundreds of millions of dollars in sales of mobile equipment, beaten out by such FM suppliers as Radio Engineering Laboratories. Sarnoff was beginning to feel the sting of being excluded from FM. It brought out the survivor in him. Like Bill Gates before the antitrust action against Microsoft, he was top

dog in what was then one of America's most powerful corporations, and he wanted to stay there. Something had to be done.

Sarnoff prepared the groundwork for what was to come in a
meeting with Franklin Roosevelt. It took place during Senator
Wheeler's investigation of the FCC rulings that had suspended
television's experimental broadcasting.[17] At this meeting, Sarnoff
revealed an intransigence that was as deep as Armstrong's, by
flatly refusing to meet privately with Fly to resolve their difficulties. The president relied on Sarnoff's political support, and
although he could not fire the FCC chief, his administration could
exert intense political pressure. Fly finally yielded and authorized
unrestricted television broadcasting over eighteen new channels
in 1941. The earlier FCC decision awarding TV's old experimental
range to FM remained unchanged for the time being, as did the
requirement that RCA use FM for television sound. Although he
was unsuccessful, a new voice, Paul Porter, CBS's legal counsel in
Washington, advocated that the 44 to 50 MHz range be returned
to television; that would have left 51 to 108 MHz for FM. Omi-
nously for FM, Porter was chosen by Roosevelt to replace Fly as
FCC commissioner in November 1944.

When the Senate asked Sarnoff how he intended to protect
RCA consumers against the danger of obsolescence in his television sets, since many features of television had not yet been firmly
established and might change radically and often, Sarnoff replied
that "we have as much interest as anyone in protecting the public.
We serve the public. They are our customers. If we disappoint
them we shall not have the business. But if this new philosophy of
obsolescence means protecting the people, then we have to go and
learn our economics all over again in this country. Many of the
things that happened in this country have happened as a result

of obsolescence—automobiles and other things. While there is a technical difference between automobiles and television, there is no difference in philosophy."[18]

These remarks reveal many things about Sarnoff, including his plans for television and the probable origins of his supplantive theory. To deal definitively with the Senate challenge over obsolescence, Sarnoff fell back on RCA's policy of refunding the difference in purchase price between sets bought in 1939 and those bought in 1940. Showing great foresight, RCA had provided each purchaser of a 1939 set with as much as a $205 refund when the price of the 1940 sets went down as a result of mass production. According to Sarnoff's testimony, this act of public service cost his company $175,000.[19] Sarnoff also minimized the technical alterations that might be required in future TVs, describing them as belonging to one or more of the following groups: "(a) the number of picture lines; (b) the number of pictures per second, and (c) synchronization." Each or all of these could be accommodated, Sarnoff claimed, by a $40 adjustment.[20]

Although FCC Chairman Fly's reallocation of the experimental 44 to 50 MHz band to FM from television prompted Senate hearings, the move actually did not make any RCA sets obsolete. The five channel tuners of the TRK 9 and TRK 12 received every frequency from about 40 to 90 MHz, and after World War II RCA issued instructions to dealers on how to retune the prewar sets to five of the current twelve channels (channels 2 through 6). This was an easy thing to do, requiring only a screwdriver adjustment.[21] After 1946, RCA's TRK 630 model received all twelve channels, rendering the earlier sets obsolete. Meanwhile, between the FCC decision in 1940 and the shutting down of commercial radio production in 1942 (due to other demands on wartime industry), licensed radio manufacturers in America produced 500,000 FM sets capable of receiving frequencies in the 44 to 50 MHz range, and Armstrong received a 2 percent royalty on every set sold.

World War II changed radio radically. For one thing, over strong resistance (possibly originating with a Sarnoff-inspired intrigue), the United States Army adopted FM as its communications standard. Somehow the Germans had already stolen FM patent secrets and used them to great advantage in their 1939 Panzer thrust into Poland. This lesson was not lost on General George Patton, whose rapid advance across France toward Germany would have been impossible if resupply and static-free field communications had not been guaranteed by FM. World War II was the first electronic war, and Armstrong—a genuine patriot—contributed heavily to the victory. Without prompting, he voluntarily suspended all of his royalty rights for the duration in order to make military communications cheaper and more effective. He also worked on Project Diana, whose mission was to adapt frequency modulation in order to extend radar detection methods. This work—some of it still classified today—included innovations in a field that would one day become stealth aviation.[22]

An unforeseen consequence of the war was to make GIs familiar with the superiority of FM technology. By 1947 four hundred FM stations were under construction in the continental United States. A new generation of men who had worked with FM in military service were now ready to advance the cause of FM and hi-fi back home. These technically savvy veterans created a new market in consumer electronics by building sound systems around their FM tuners, with separate components assembled from different manufacturers. Columbia Records responded by adopting improved radio transcription techniques in their long-playing records—soon to become a central player in the hi-fi boom. Small companies like Altec, Bogen, Browning, Fisher, and Pickering, which rose to prominence by making sound-system components, could not have existed without Armstrong's FM inventions.

Sarnoff had a very good war, too. A much more canny businessman than Armstrong, he went against the tide of companies providing the military with free patent licenses to aid the war effort. On the advice of Charles Jolliffe, now RCA's chief engineer, the company resisted government requests for free use of its patents, and RCA eventually won an annual fee of $4 million for the United States' use of proprietary patents during the war.

Sarnoff's next move was to arrange a commission for himself. Since he was already a colonel in the Signal Corps Reserve, this was relatively easy. In 1942 he became head of the U.S. Army Signal Corps Advisory Council. Acting on Sarnoff's recommendations, the entire command, control, and communications structure of the Signal Corps was streamlined until it began to operate with wartime efficiency. He was so successful that the chief signal officer recommended Sarnoff for an important role in planning D-Day. Eisenhower, then supreme commander of U.S. armed forces, chose Sarnoff to coordinate radio and print communications prior to the Allied invasion of Europe. Sarnoff developed a personal relationship with the future president and also spent time with Churchill and his wife. Already a powerful man before the war, Sarnoff acquired considerable personal influence from his role in planning the Normandy Invasion.

Back in the states, the ambitious Charles Jolliffe had been left in charge of RCA. He had maneuvered the FCC into calling on the radio industry to review its postwar frequency needs in preparation for hearings that would set future standards and assign future bandwidth. But Jolliffe was suddenly stricken with debilitating intestinal cancer, and by October 1944 RCA was drifting and directionless. Sarnoff dropped everything and returned stateside, taking up his corporate duties with renewed energy. By November he had engineered Paul Porter's appointment to the position of FCC commissioner in time for the hearings to begin. In December he was promoted to brigadier general, and thereafter all RCA employees were required to refer to him as "The General." One obvious implication of this new title was that he now outranked the man known at RCA as "The Major," Edwin Howard Armstrong.

The 1944–45 FCC hearings were a marvel of deception orchestrated by Porter and Sarnoff. NBC and CBS put forward a joint proposal that FM now be moved for its "own good" away from the lower frequencies it had occupied since 1940, in order to prevent sunspot interference from disrupting the fidelity of its broadcasts. Nearly sixty years have passed since these hearings, and it is impossible to know with certainty the author of the sunspot strategy, but it was both politically brilliant and technically acute—two facts that limit the field considerably.

At that time, sunspots (or "ionospheric interference") were newly discovered and poorly understood, even by experts. The legal minds of the FCC would have been hard pressed to follow the technical arguments of network engineers, even if they had wanted to do so. Apparently they did not. A former FCC engineer, one of Sarnoff's old Signal Corps cronies, Kenneth A. Norton, was accepted as an objective source. He testified that FM should be moved in order to prevent trouble from this hypothetical interference—interference of a kind that had never bothered FM broadcasting during the six years it had been on the air. Porter's FCC now had its pretext, and it moved all of the FM transmission bands from their previous position in the 44 to 50 MHz range to a new spot at 88 to 106 (later 108) MHz. This meant that the 500,000 FM radios manufactured in America before the war could no longer receive FM broadcasts. The new frequencies assigned to FM stations were simply beyond the range of prewar FM channel selectors.

Television—somehow strangely immune from sunspot interference—was then reassigned to the 44 to 88 MHz range, which conveniently made the receptive range of the TRK 9 and TRK 12 audio tuners once more fully operational. It was a total victory for Sarnoff and for RCA. Unfortunately for Armstrong and FM, this cynical and calculated FCC decision made half a million FM tuners, all of those built between 1940 and 1945, inoperable. These tuners were just not built to access new frequencies so far removed in the spectrum from their old position. This clever underhanded action by the FCC impeded the progress of FM broadcasting in America for years. Unlike Europe, where several countries created national FM networks shortly after World War II, Sarnoff's obstructive efforts prevented America from developing an FM network until National Public Radio began broadcasting in April 1971.

The FCC ruling raised a public stink, of course, and another Senate investigation.[23] RCA and the networks simply rode out the storm. Armstrong tried every form of redress before filing suit against RCA, but it was now 1948, and Bell Labs had just unveiled the first germanium point-contact transistor. Radio was about to change again, forever. Armstrong had only five more years left on his basic FM patents, so RCA's lawyers simply dragged out the litigation until it broke Armstrong, financially and spiritually.

He was a much different man by the 1950s. Frustrated, bitter, and jealous, he fought with Marion over money and over her attention. By 1953 his FM patents had expired and his scientific legacy was in doubt. On Thanksgiving Day, Marion left him, after he struck her on the arm with a poker when she refused to give him access to the $750,000 they had set aside for their retirement. Armstrong spent Christmas and New Year's Eve alone for the first time since their marriage twenty-one years before, and he could not face the prospect of Valentine's Day without her. On January 31, 1954, he jumped to his death from the balcony of their Manhattan apartment, leaving her a two-page letter expressing "how deeply and bitterly I regret what has happened to us." RCA executives attended Armstrong's funeral en masse and, perhaps genuinely, Sarnoff cried demonstrably. Later that year, RCA settled with Marion's lawyers for the same $1 million without royalties that Sarnoff had first offered Armstrong in 1940.

Eventually Sarnoff became an American icon, but his influence at the FCC declined steadily following Armstrong's tragic death. A sea change in the commission's sense of public responsibility dates from this era. When Sarnoff asked his friend Dwight Eisenhower to be promoted to major general in 1956, Eisenhower demurred. Sarnoff then lobbied the Senate Armed Services Committee to award him the Distinguished Service Medal, the third highest military honor (World War I recipients included Marshal Foch, General Petain, and General Pershing). The committee politely refused. Later, when Sarnoff set his sights on the Presidential Medal of Freedom, America's highest civilian award and an honor that Lyndon Johnson had just provided posthumously to Jack Kennedy, Johnson also sidestepped him. Undaunted, Sarnoff decided the time had come to share his greatness with the world. In 1965 he hired his cousin, Eugene Lyons, to write the biography that would appear on his seventy-fifth birthday in 1966. Unfortunately, the first draft cast him in a bad light, remarking on his egotism, his insatiable womanizing, and his tendency to have affairs with friends' wives. Sarnoff characteristically took control of Lyons's manuscript, something anticipated by his contract. He rewrote it as the glowing third-person hagiography we have today.

Here's what the book says about Marion Armstrong: "Sarnoff's secretary was a tall, strikingly handsome girl, Esther Marion MacInnis. The Inventor was smitten with her at once and laid a long siege for her affections. Though gangling and prematurely bald, Armstrong was a magnetic personality. But he had eccentricities that gave a sensible New England girl pause. One of them was compulsive speeding in the most high-powered cars he could find at home or import from Europe."[24]

In the 1950s, after winning settlements in her husband's lawsuits and campaigning to restore his reputation internationally, Marion Armstrong spent considerable effort locating his Hispano-Suiza, which she fully restored and repainted a deep, cobalt blue. Until her death in 1979 she drove it every summer along the ocean road in Rye Beach, New Hampshire, her broad-brimmed hat twisting and flapping in the wind. Even as an aging widow, she drove very fast—that was the way she and Howard had always liked it.

MINIATURIZATION

The 1950s marriage of transistors and printed circuits rendered the postwar generation of consumer electronics obsolete, because it made them, for all intents and purposes, unrepairable. Printed circuits in the mass-consumed transistor radios that began to flood the American market from Japan after 1957 contained machine-soldered parts that were too small to permit easy replacement, and too inexpensive to make service practical. The casings of these radios reflected their disposability. The bright, brittle plastic of Japanese transistor radios afforded no real protection against breakage. Designed for the teenage market, these conspicuously consumed radios changed color and style with every shipload.

Before transistors arrived on the scene, manufacturers liked vacuum tubes because they too encouraged repetitive consumption. They burned out like flashlight bulbs and had to be replaced. Even the smallest tubes were held in place by specialized sockets that permitted easy repair. Whenever a radio went dead, its owner unplugged the set, opened the back, and detached one or several tubes, using a pair of specially made rubber-tipped tongs (ubiquitous in the 1960s). He then put the suspect tube(s) into a paper bag and went to a department store looking for a special machine called a tube tester, which identified the burned-out tube. Armed with that information, he marched over to the tube department and purchased a replacement, which he installed in the back of his radio.

The whole process was not much more complicated than replacing a light bulb. It saved the expense of hiring a technician to make a house call, and it avoided the inconvenience of transporting a cumbersome, old-fashioned radio receiver to the shop. It also involved consumers in the process of servicing their own radios, which encouraged a widespread awareness of radio issues among a self-educated consumer universe. This is always very good for business, since it makes high-end customers more susceptible to the most recent technical improvements. The size of receivers containing vacuum tubes prior to miniaturization encouraged public involvement in "radio art" on a scale that has no point of comparison today (although marketing specialists still acknowledge that product education paves the way for increased consumption among the wealthiest consumers).

Miniaturization began as a cost-cutting strategy during the Depression. Out of necessity, the size and price of radio sets were reduced suddenly following the crash of 1929. Highboy receivers, with their large speakers and resonant cabinets, were simply unaffordable in the 1930s. Their extravagant construction rendered them effectively obsolete. An assortment of smaller upstart companies in outlying areas like Los Angeles began to manufacture radios that sat on top of the kitchen table or sideboard. The cabinets of these smaller radios had names like the Tombstone or the Cathedral, but generally these Depression-era tabletop radios were called "midgets," and sold in the range of $35 to $55, much cheaper than the $200 asking price of their 1920s counterparts. The full-throated timbre of the Highboy gave way to nasality in its smaller cousins. But the vacuum tubes these midgets contained were more or less the same size as those in the pre-Depression models. Midgets simply used fewer tubes.[25]

In 1936 Hivac, an English vacuum tube company, began producing very small tubes for use in hearing aids. By chance, Norman Krim, who had been one of Vannevar Bush's graduate students at MIT and was then working at Raytheon, found out about Hivac's innovation. He surveyed American hearing aid manufacturers and found a potential domestic market for similarly small, low-drain tubes. In 1939, for a tiny investment (and a promise to Raytheon president Laurence Marshall that he would resign if he was wrong), Krim produced the CK501X. Raytheon quickly made a profit by producing smaller hearing aids than were ever before possible.[26] The first of these were manufactured by Sylvania. Although this innovation was very good news for the hearing impaired, the most important feature of the smaller vacuum tube was its timing. At the beginning of World War II, Krim's subminiature tubes had very important military implications.

In Europe around 1935, a brilliant Jewish doctoral student of electrical engineering could not find work in a rapidly Nazifying Austria. During his ample spare time, he built himself a radio so that he could hear more objective international news reports while saving money. Paul Eisler also became fascinated with photogravure engraving and conceived the idea of printing circuits onto copper plates instead of wiring them in standard point-to-point construction. To test his idea, he took his lovingly constructed radio apart and made a wiring pattern on copper foil. He backed this with paper and glued the circuit to a flat sheet of salvaged Bakelite before attaching vacuum tube sockets. When it was done, he insulated the exposed copper with varnish. This cheap, compact radio worked very well. Eisler could receive Schweitzerdeutsch stations easily from his Vienna apartment.

In time, Dr. Eisler fled Vienna and presented his prototype printed circuit board to a radio manufacturer in London, who did not buy it. Unbothered, Eisler sold another idea to Marconi Wireless for a small amount of money and established a workshop in his tiny London apartment. He began to make a living as a freelance electrical technician and inventor and was so successful that he could afford to send for his entire family well before Germany annexed Austria in 1938. Two years later, during the early days of the London Blitz, it became obvious that anti-aircraft fire was ineffective because the shells would not explode until they made contact with aircraft. Hitting a German bomber directly in midflight was a tall order, so Eisler pushed the idea of using his printed circuits to permit the miniaturization needed for proximity fuses. A proximity fuse can detonate artillery shells that are simply *close* to a target.

It now seems genuinely unclear whether the very first effective radio proximity fuse was a British or an American triumph. In either case, Eisler got in on the ground floor. The proximity fuse that went into service in the Pacific and European theaters in 1944 was immeasurably more effective than any contact detonator.[27] Its military significance is indicated by the fact that the NKVD used Julius Rosenberg's network to infiltrate Emerson Radio and steal proximity fuse blueprints and spare parts.[28] By the war's end, over four thousand German V1 rockets had been destroyed by artillery shells armed with proximity detonators. Printed circuits were here to stay, and their association with an ethic of disposability (fuses are used only once, after all) had become firmly established.

Back in America, Raytheon was now rich. Before the war, the company had registered a respectable $3 million in sales annually, but Krim's subminiature vacuum tubes had been used in proximity fuses for every anti-aircraft shell fired in 1944 and 1945. War contracts had increased business to $173 million per year. In a brainstorming session among Raytheon executives in early 1945, Krim suggested that Raytheon produce a miniature pocket-size AM radio that would use a hearing aid earphone instead of a speaker. Once again, the initial investment he required was relatively small, only $50,000, or about twice as much as the subminiature tube project he had proposed in 1938. In view of the fact that his first idea had yielded a 6,800 percent increase in company sales, Krim was given the green light.

Raytheon expected their new pocket radio to spearhead a whole line of electronic consumer products, and in anticipation of the manufacturing demands, the company bought Belmont

Radio Corporation in 1945. The first manufactured pocket radio, the Belmont Boulevard, debuted in December 1945 with a full-page ad in *Life* magazine. It was a sweetly designed, perfectly functional pocket radio that is now so highly prized by radio collectors that it is considered virtually unavailable. In recent years there have been only two known sales: one sold in the early 1990s for $750; and another, in near-mint condition, sold on eBay in 2001 for $1,680. Each of the Boulevard's five subminiature vacuum tubes is held in a socket from which it can be detached and replaced for repair.[29]

Despite the interest and excitement the Boulevard caused as a curio, few Americans were willing to spend $30 to $65 for a miniature radio, and only about five thousand were ever purchased. This represented a marketing disaster for Raytheon, and it contributed to image problems for pocket-sized radios for years to come. Because of the negative publicity, details surrounding the Boulevard's creation were omitted from Raytheon's authorized corporate history, *The Creative Ordeal*. Still, the idea of a workable pocket radio did not die. It simply went underground. Kit manufacturers like Pocket-Mite (1948) and Private-Ear (1951) kept the genre alive as a cult phenomenon for techies.

Two years after the Boulevard's debut, two young researchers at Bell Labs in New Jersey worked up a replacement for the vacuum tube. In late 1947, using a piece of germanium crystal, some plastic, and a bit of gold foil, John Bardeen and Walter Brattain managed to boost an electrical signal "almost a hundredfold."[30] Strangely, their point-contact transistor met with hostility and envy on the part of their boss, William Shockley, who immediately set about trying to improve their invention. And as a matter of fact, his junction transistor made the amplification of electric signals more manageable than Bardeen and Brattain's original "ungainly invention."[31] Despite these differences, all three men shared the 1956 Nobel Prize for Physics as co-inventors of the transistor.

Western Electric held the patents on Shockley's junction transistor. In the early 1950s the company licensed transistor production at several American and one Japanese firm. Of these, Texas Instruments and the Tokyo-based company that would become Sony are most significant in the history of obsolescence. In 1953, one year after purchasing its patent license from Western Electric for $25,000, TI beat all other American companies in manufacturing a silicon junction transistor. By 1954 Texas Instruments returned to what had now become the Holy Grail of miniaturization for techies and electricity buffs. Their shirt-pocket radio hit the market in October 1954, just in time for Christmas purchases. Produced in partnership with IDEA, an Indianapolis electronics firm, the four-transistor Regency TR-1 was available in a variety of colors (my personal favorite is red). It had bad tone quality and—even though TI lost money—it still cost a hefty $49.95. Nonetheless, it was the world's first transistor radio, and IBM's president, Thomas J. Watson, was paying attention.

In 1955 Watson purchased over a hundred of the TR-1s and kept them in his office. Whenever executive discussions came around to vacuum tube technology and obsolescence, Watson would make his point by taking one of the variously colored Regency TR-1s out of his desk drawer and presenting it to whoever did not yet own one. By 1957 Watson had signed Texas Instruments to supply silicon junction transistors for the first fully transistorized IBM computer, the 1401, completed in 1959. But when Seymour Cray at CDC beat IBM to the punch by producing the CDC 1604 in 1958, Watson instituted a policy that all vacuum tubes be eliminated henceforth from IBM design and manufacturing.[32] The silicon junction transistor had superseded the vacuum tube by eliminating its characteristic inefficiencies—warmup time, bulkiness, high power drain, and fragility. Although hybrid electronic products (especially manufactured radios from Emerson and Automatic) continued to combine transistors and vacuum tubes well into the 1960s, manufacturers increasingly dumped their tube inventories. Clearly, by the time of Watson's decision in 1958, tube technology was on its way out.[33]

No one knew this better than Masaru Ibuka and his partner Akio Morita, who had purchased a license to manufacture transistors from Western Electric in 1953, paying the same hefty $25,000 fee as Texas Instruments. They were nearly two years behind TI in transistor production when they got the license, so it is remarkable that Sony was able to produce its first transistor radio (the TR-55) for the domestic Japanese market in 1955 only one year after the Regency TR-1. That same year, Zenith—a name literally synonymous with the highest quality in radio—introduced a high-end transistor, the Royal 500 or Owl Eyes transistor radio (so called because the round tuner and volume controls resembled large eyes). The original Royal 500s were very effective receivers that used seven transistors in a hand-wired point-to-point circuit. But when the hand-wiring required to produce this pocket radio sent Zenith's costs through the roof, they switched to circuit boards. This point was not lost on Sony.

In 1957—the first year that Dick Clark's American Bandstand appeared on national television—Sony entered the American market with the TR-63. These almost affordable pocket transistors allowed a new generation of music listeners to rock around the clock with their new music wherever they went. But the transistors in Sony's TR-63 were still hand-wired. The connections were threaded through holes in a circuit board and then hand-soldered by inexpensive nonunion Japanese laborers. As early as 1952, an alternative to hand-soldering had been developed.[34] The way that Motorola competed with the low production costs of Japanese firms was by automating its soldering process, using a single controlled dip.

This technique reduced Motorola's wiring costs and its size requirements enormously, resulting in even smaller transistor radios. But it also transformed the pocket radio into an unrepairable product whose lifespan was limited by the durability of its original components. Pocket radios were now disposable products because their circuit boards were too small to be repaired by hand.

Of course, the actual durability of the parts used in the radio could also be controlled with frightening accuracy by scientific product and materials testing. By the 1950s, product life spans were no longer left to chance but were created by plan, and it is at this moment (from about 1957 on) that the phrase planned obsolescence acquired the additional meaning of "death dating." Eventually, death dating would become the primary meaning of the phrase.

The people of the United States have been thrust into making a more abrupt transformation in their system of values since WW II than in just about any comparable period of time in the nation's history . . . Some of the changes . . . are directly related to the pressures and stimulations that encourage Americans to increase our consumption.

VANCE PACKARD, *THE WASTE MAKERS* (1960)

5   The War and Postwar Progress

Even before it had a name, nylon made two natural fibers obsolete: hog's hair and silk. In the 1930s the principal use of hog bristle was in the manufacture of brushes—for painting, washing bottles, stimulating the hair and scalp, and cleaning one's teeth. The market for bristle brushes was relatively small, however: not everyone brushed his teeth, or even had teeth. Among those who did, our modern twice-a-day habit of tooth brushing was not entrenched as it is now.

Silk, on the other hand, was big business—$100 million in annual imports to the United States alone. With the radical shortening of skirts in the 1920s, American women became very conscious of how their legs looked, and as the decade roared on they spent more and more money on expensive, full-fashioned, often flesh-colored silk stockings. But by the second year of the Depression, silk hosiery was an unaffordable luxury for most American women. The only economical alternative, rayon stockings, were too coarse, too shiny, and too saggy. What was needed was a radical innovation in the development of synthetic silk.

THE SILK TRADE

Japan, the nation whose superior silk constituted 90 percent of the fabric imported by America, was also suffering from the American Depression. As much as 25 percent of all Japanese workers were employed in some aspect of silk manufacturing. Although Japan was the only industrialized country in Asia, it had few indigenous natural resources and therefore relied heavily on the hard foreign currency it obtained through the silk trade to import aluminum, iron, rubber, and other requirements of industry. Fewer American silk sales meant less money to buy these raw materials and therefore fewer manufactured goods to sell on its domestic and foreign markets.

By 1931 the Japanese were desperate. Their American foreign trade had dried up overnight, and across the Sea of Japan to the west the Kuomintang were giving every indication that they would eventually unify China and kick out the foreign powers who were exploiting its natural resources. Japan was one of these powers. At that time the Japanese Kwantung army was the main military presence in Manchuria, one of Japan's main suppliers of cheap raw materials. After a bomb of unknown origin exploded near Japanese troops guarding the train station at Mukden (Shenyang), the Japanese invaded all of Manchuria and instituted martial law. Despite a promise to the League of Nations to end their occupation and return to the railway zone, the Japanese set up a puppet government under Henry Pu Yi, the "last emperor" of China. Manchuria (renamed Manchukuo) was recognized internationally by only three of Japan's allies and trading partners: Germany, Italy, and Spain.

As if eager to confirm the world's worst fears, the Japanese, in their desperation for foreign currency, took over and expanded the cocaine, opium, heroin, and morphine trade up and down the Chinese coast. They also used opium to control Manchukuo's puppet emperor, whose wife was hopelessly addicted. During the 1930s the Japanese government earned over $300 million a year from distributing narcotics.[1] Simultaneously, an abundance of inexpensive Manchurian "cotton morphine" became available to California drug rings.[2] By the mid-1930s few Americans imagined that their opinion of Japan could get much worse.

Already a formidable naval power in the Pacific, Japan began to advocate what it called a Great East Asia Co-Prosperity Sphere, in direct opposition to the goals of the American-backed Kuomintang, and in 1933 Japan forced the Kuomintang to establish a demilitarized zone on one of Manchukuo's borders.[3] In the meantime, Japanese silk exports to the United States had begun to recover from their early Depression low, and it rankled some patriotic Americans that U.S. dollars were supporting Japan's blatantly militarist ambitions in China. Suddenly, the prospect of developing a viable synthetic substitute for silk took on strategic and political significance.

American strategists in Washington began to speculate about the effectiveness of using a trade embargo to end Japanese expansionism. Japanese industry was especially vulnerable to embargo because it relied exclusively on imported raw materials. From 65 to 100 percent of Japan's supplies of aluminum, cotton, iron, lead, mercury, mica, molybdenum, nickel, oil, rubber, and tin were imported, much of it from the United States itself. If America imposed a trade embargo on Japan, in retaliation the Japanese would cut off America's silk supply.[4] But ironically, this would actually make an embargo *more* effective, since Japan would lose a major source of the foreign currency its industrial economy needed to survive. And if an effective silk substitute could be created, political pressure to end the embargo prematurely would not build. America could then double its economic benefits, by developing a profitable indigenous industry *and* by retaining the $100 million it paid out annually to a foreign power.

A lot was riding on the venture to produce synthetic silk, and the chemical firms of many nations eagerly entered the competition. Although it had not yet been invented, artificial silk was clearly the fabric of the future.

BETTER LIVING THROUGH CHEMISTRY

In America, the mammoth political and economic storm gathering over the Pacific forced DuPont to concentrate its research energies on producing a synthetic replacement for silk. Already a successful multimillion-dollar corporation, DuPont had strategic and political problems of its own. In the early 1930s the company's best-known product was TNT, and family members had been vilified in the media as being "the American Krupps" for supposedly profiteering with TNT production in World War I, just as the German firm Krupp had profiteered with its howitzer field and anti-aircraft gun. In 1934 the U.S. Senate conducted a hearing in which senators questioned DuPont's owners about the jump in annual profits from $5 million to $60 million during the war. In fairness, it should be said that DuPont's domestic products greatly outnumbered anything they produced for the military. But DuPont's success had attracted interest, and the family was vulnerable because its public image was that of a manufacturer of weapon-grade dynamite. Desperately in need of a corporate makeover, the company began to survey advertising agencies.

In 1935 DuPont hired Bruce Barton of Batten, Barton, Durstine & Osborn, a premier advertising firm. Barton was an expert at image changes. In 1925 he had written a book detailing how Jesus had actually been one of the very first ad men. Without any irony at all, Barton explained that "as a profession, advertising is young. As a force, it is as old as the world."[5] For $650,000 Barton succeeded in transforming DuPont into a warmer, family-friendly company, one that created "Better Things for Better Living . . . Through Chemistry." Along with this new publicity campaign, DuPont initiated an aggressive program to develop domestic products. This was the moment when Elmer Bolton decided to dedicate the company to creating a commercially viable synthetic silk: "The tremendous advance in the artificial silk industry in recent years has emphasized the importance of developing an entirely new textile fiber, in our opinion [this is] one of the most important speculative problems facing the chemist today."[6]

DuPont had been experimenting in the new theoretical field of polymers for some time. Back in 1927, Bolton had hired a gifted and prolific lecturer in chemistry from Harvard, Wallace Hume Carothers, to head up its Pure Science Division, one of the first research teams in American industry. Carothers had the idea of building up "some very large molecules by simple and definite reactions in such a way that there would be no doubt as to their structure."[7] Tired of academic penny-pinching, Carothers hired an expensive eight-member research team that included recent chemistry PhDs, and he set about joyously spending money on the laboratory (nicknamed Purity Hall) and the equipment they required.

Early on, the DuPont team developed neoprene, a synthetic rubber that would later have widespread commercial applications. But 1934 was the year of the company's breakthrough in the synthetic silk competition. In a whirlwind of inventiveness, Carothers had solved the basic problem of polyamides. Around New Year's Day of that year, Carothers wrote to a close friend describing an intense period of creativity he had just experienced: "At the time your letter arrived my head was practically on fire with theories. I had to write 6 new projects and I stuffed them full of theory. The prize one is this."[8] What followed was a detailed description of the polymerization of acetylenes, the essential missing step in creating the large chains of molecules necessary for a new commercial fiber.

Unfortunately, Carothers suffered from bipolar disorder before lithium salts made the disease controllable. Although his biographer wrote that Carothers did not experience the manic highs that many people with this disorder report, Carothers did alternate between periods of great creativity and deep depressions during which he would often disappear for days, either to medicate himself with alcohol or retreat to a psychiatric clinic in Baltimore. In the last years of his life Carothers carried with him a container of cyanide. His closest friends understood and tolerated this as his guarantee of having the ultimate escape.

By May of 1934 Carothers was once again intensely depressed. This time he disappeared into the Pinel Clinic near Johns Hopkins University, where his friend Julian Hill eventually found him. Meanwhile, back at the DuPont lab, Don Coffman worked from Carothers's notes to produce a polymer with high molecular weight. Coffman immersed a "cold stirring rod in the molten mass [and obtained] upon withdrawal a fine filament . . . It seemed fairly tough, not at all brittle and could be drawn to give a lustrous filament."[9] This polyamide fiber became known as Fiber 66.

Within days Carothers was back on the job, and during a year of experimentation that followed, he discovered that Fiber 66 was much stronger than silk and would not dissolve in dry cleaning solvent. Its 200° C melting temperature was also far above that of organic fabrics. But the coup de grâce was that both of the six-carbon monomers making up this polymer could be manufactured from a single chemical source, benzene. Although Carothers himself favored the commercialization of another polymer his lab had developed (Fiber 5-10), Bolton settled on Fiber 66 because of its extraordinary properties and the fact that it required only a single, readily available raw material, crude oil. DuPont's Pure Science Division spent 1936 developing methods of spinning the new fiber into threads that could be woven into fabric.

Wallace Carothers would not be part of this phase of R&D at DuPont. In January of that year he received the tragic news that his beloved younger sister, Isobel, thirty-six years old, had died. Isobel was a radio personality whose daily sketch *Clara, Lu 'n Em* was nationally syndicated on the Blue Network and is today regarded as the first-ever broadcast soap opera. During an especially cold and damp winter in her native Chicago, at a time when penicillin was still an experimental drug and largely unavailable, Isobel developed pneumonia. Wallace Carothers suffered a complete mental breakdown when he learned of her death.[10]

Although he married Helen Sweetman the following January, Carothers never really recovered. In April 1937 he finally succumbed to the depression that had plagued him all his life. A year before Fiber 66 made its debut as nylon, and two years before the first nylon stockings were produced, Wallace Carothers checked into a Philadelphia hotel, swallowed the contents of his bottle of cyanide, and collapsed. He had just turned forty-one.

After DuPont learned that Helen was pregnant with Wallace's child, the company arranged a series of bonus and royalty payments for the young widow and her new daughter. Later, when Carothers's estranged parents suffered financial hardship due to age and unforeseen medical expenses, Elmer Bolton, Carothers's

old boss, persuaded DuPont to provide them with a modest
monthly stipend until the end of their days.

Tragedy was everywhere in 1937. Japan's allies, Germany and
Italy, created the Axis alliance and jointly sent reinforcements to
Spain. Two days before Carothers's death, the Germans bombed
the loyalist town of Guernica out of existence, in a beta test of the
aerial bombardment techniques they would use in World War II.
In June of that year, Stalin began his infamous purge of the Red
Army, an atrocity that would leave 20,000 Russian soldiers dead.
In July, Amelia Earhart disappeared forever into the airspace over
New Guinea. And, after landing in the seaport of Tianjin, Japanese
invasion forces took China's northern capital, Peking (Beijing),
before moving south to capture Shanghai, where they encountered substantial resistance for the first time. With Shanghai
eventually subdued, in early December 1937 the Japanese invaded
China's southern capital (Nanjing) in a viciously punitive action
now called the Rape of Nanking. Perhaps as many as 300,000 ci-
vilians died at the hands of Japan's poorly disciplined and badly
trained recruits.

On December 12, one day before Nanking fell, the Japanese
bombed and sank the *USS Panay,* a Yangtze River gunboat, in a
deliberate act of provocation. America readied itself for war. But
when the Japanese apologized and made restitution for their
"mistake," Roosevelt accepted the apology. The United States
breathed a collective sigh of relief, even though war with Japan
now looked inevitable. In America, four years before Pearl Harbor, hostility toward the Japanese had reached a new high. By
mid-December a boycott that began when the League of Nations meekly voted their "moral support" for China garnered the
backing of over fifty American hosiery companies, who voluntarily shifted from manufacturing silk stockings to manufacturing

stockings made with silk substitutes. In 1937, to American eyes, even coarse, baggy rayon looked better on women's legs than Japanese silk.

A heady brew of fear and hostility fired America's imagination about Japan. Less than a year after the horrific radio reports of the attack on Nanking, Orson Welles's broadcast of *War of the Worlds* on October 30, 1938, sent the United States into an overnight panic. This scare allowed the *Philadelphia Record* to run a story on November 10 headlined: "DuPont's New Fiber More Revolutionary than Martian Attack." The unnamed fiber was "so incredible" that the implications "could barely be summarized." One of these implications was the deathblow that America could now deal to the oriental silk trade, an act that would deliver more strategic damage to Japan than the sinking of Hirohito's navy.

Articles like "Is Nylon's Future the Past of Silk?" so frequently repeated the promise that nylon would soon make Japanese silk obsolete that it became accepted wisdom, even though the word "obsolete" was rarely used.[11] A front-page story in the *New York Times* on October 21 concerning a new synthetic silk factory ran with this headline: "$10,000,000 Plant To Make Synthetic Yarn: Major Blow to Japan's Silk Trade." The article noted that hosiery for women "has remained almost exclusively an outlet for raw silk because synthetic yarns produced up to now have been too lustrous, too inelastic and insufficiently sheer for production of hosiery for the American market . . . [But the] introduction of these new yarns . . . would eat deeply into the market for silk hosiery. The DuPont yarn . . . can compete with all grades of silk stockings retailing at $1 a pair or less . . . It could compete with much more expensive silk hosiery."[12]

At BBD&O, Bruce Barton made sure that DuPont capitalized on this valuable anti-Japanese publicity to improve its own image

and enlarge the future market for synthetic silk hosiery. American
women were writing daily to DuPont, angrily refusing to spend
money on Japanese silk stockings until the Kwantung army withdrew from China.[13] DuPont's makeover was taking hold, and its
slogan "Better Living . . . Through Chemistry" was becoming a
beacon of hope for patriotic Americans. Children's home chemistry sets appeared under Christmas trees across America.

The first Fiber 66 product went on sale in 1938, in the form of
Dr. West's Miracle Tuft Toothbrushes. Bristles in these brushes
lasted three times longer than hog's hair. Still, DuPont's miracle
product did not yet have a proper name. Generations before Ira
Bachrach founded the NameLab, a San Francisco development
and testing firm that specializes in creating names for new products and companies, the word "nylon" became one of the first deliberately developed and tested consumer product names. Like
most NameLab products (including NameLab itself), DuPont's
list of candidates was based on combinations of meaningful word
pieces. This fact probably explains how the urban legend developed that "ny-lon" was a conflation of "New York (NY)" and
"London." Actually, over four hundred candidates were focus-
group tested for the consumer's response, including Amidarn,
Artex, Dusilk, Dulon, Linex, Lastica, Morsheen, Norun, Novasilk,
Nurayon, Nusilk, Rames, Silpon, Self, Tensheer, and Terikon.
Among these, Norun was the frontrunner among the women
surveyed, for obvious reasons, but the experts concluded that it
sounded too much like neuron or moron and would invite tasteless jokes. Dr. Ernest Gladding, chairman of the name committee
and later director of DuPont's Nylon Division, experimented with
different vowel and consonant combinations before settling on
nylon because of its mellifluity. The word officially replaced Fiber

66 in early October 1938, two weeks before the new material's public debut.[14]

After conducting more experiments in late 1938, DuPont sold its first nylon stockings to the company's female employees on February 20, 1939. These women were allowed to buy only two pairs of stockings in a single shade for $1.15 each. As a condition of sale, the women had to complete a questionnaire about their nylons within ten days. Seven months later, on October 30, 1939 —in a month when German invaders had annexed western Poland, and President Roosevelt had received a letter, signed by Albert Einstein, urging the United States to develop the atomic bomb—stockings went on sale to local customers in Wilmington, Delaware, and four thousand pairs were sold in three hours.

These early stockings were much thicker and more durable than the progressively finer nylons produced after the war. Some second-hand accounts support the claim that DuPont reduced the thickness of their nylons in order to make them less durable, although no one has evidence to prove this.[15] What can be easily demonstrated is that throughout its long history of nylon production and marketing, DuPont has been keenly aware of its product's life cycle (PLC) and the role of psychological obsolescence in generating repetitive consumption. For example, in the 1960s, when acceptance of the bare-leg look threatened to put nylons out of style, DuPont responded with tinted, patterned, and textured stockings, rendering obsolete the old-fashioned notion that nylons come only in skin tones. This allowed DuPont not only to secure its market but to expand its sales by introducing yearly fashion changes.[16]

By the time nylon stockings became nationally available on N-Day, May 15, 1940, intense demand had already been created through a cleverly orchestrated marketing campaign emphasizing voluntary silk deprivation, anti-Japanese patriotism, and product superiority. Across the nation, many women knew about the thermoplastic properties of the new stockings long before they had actually seen a nylon. On N-Day they lined up by the thousands, desperate to buy DuPont's new miracle hose. Never before had any consumer product enjoyed such immediate success. Papers ran headlines like "Nylon Customers Swamp Counters as Opening Gun Is Fired" and "Battle of Nylons Fought in Chicago." In the product's first six months on the market, 36,000,000 pairs of nylons were produced and sold. In 1941 the number rose to 102,000,000 pairs.[17]

The popularity of nylons was immediate and complete. By 1943, across America's airwaves, Nat King Cole used the nationwide scarcity of nylons to ask forgiveness of a wartime sweetheart: "Baby, let bygones be bygones / 'cause men are scarce as nylons." Meanwhile, a Russian science and technology espionage team in New York paid $500 to an American informer they called Khvat (the Vulture) to steal DuPont's patented secrets and pass them along to the Soviets.[18] Everyone was clamoring for the new silk substitute.

Because 1940 was an election year, Roosevelt saved the option of a formal trade embargo on Japan until weeks before the elections. On September 26 he banned the sale and shipment of all scrap iron to Japan. The Japanese responded immediately by formally joining the Axis, an act intended to threaten America with a two-front war if it continued to oppose Japanese expansion. But Japan's formal alliance with Germany and Italy met with outrage in America, and it solidified Roosevelt's support. With public opinion now firmly in favor of punishing Japan, Roosevelt stepped up the trade embargo in the following months, to restrict the flow of all goods. By August 1941 the Pacific Common-wealth countries, the Netherland Indies, and the Philippines joined the U.S. effort to cut off Japan's supply of commercial and industrial materials, as well as imported rice and other basic food-stuffs.[19]

Days before the Pearl Harbor attack, the *Wall Street Journal* ran a story called "Japan: Its Industries Live on Borrowed Time," iden-tifying Japan's depressed silk industry as the backbone of her economy.[20] On the very day of the attack, the *New York Times* re-ported that the embargo had "forced a curtailment of perhaps 40% in Nippon's factory operations."[21] From the Japanese per-spective, the preemptive Pearl Harbor attack may actually have been a strategic masterstroke intended to end America's ability to enforce the embargo while simultaneously exposing the nations of the Pacific rim to further Japanese expansion. Unfortunately for Japan, it strengthened America's resolve and won a meager six months of victories before the decisive Battle of Midway.

The United States did not formally declare war until February 2, 1942. At that time, DuPont, which had done so well with mili-tary contracts in World War I, turned all of its nylon production over to the war effort. The company also made a special nylon flag that flew over the White House for the duration of the conflict. It symbolized America's creative industrial capability as well as its determination to do without Japanese silk.

Michael Schiffer, an anthropologist, archaeologist, and histo-rian of material culture, has objected to the way that technological competitions of the recent past are framed "in terms of a seem-ingly inevitable winner and loser."[22] And in the case of silk, he is certainly right. Even though nylon had a number of advantages over its organic predecessor, it was not destined to supersede silk entirely. The contest between silk and nylon is not really the story of a superior technological innovation replacing an inferior natural product. It is the story of a symbolic contest between two cultures fighting for economic dominance. America's eventual victory in World War II did not make silk disappear. Silk simply lost the power it once had to industrialize a feudal country and turn it into a major international player within the space of three generations.

Even today, silk stockings can be purchased around the world. But after the development of nylon, they became exclusively a luxury item, partly because silk is more expensive to produce than its synthetic substitutes, and partly because silk stockings have a complicated history of cultural associations with luxury and privilege stretching back five thousand years. What is truly obsolete today is the war era's understanding of silk stockings. To us, silk hosiery no longer represents the threat of a hostile foreign power bent on ruthless domination of Asia. Indeed, because silk in America is primarily used in the manufacture of lingerie and sheets, we associate it with gratifications of the most intimate kind, not with geopolitical struggle.

SUBURBAN OBSOLESCENCE

Prior to the Depression, the housing industry employed 30 percent of Americans. But in 1929 it collapsed. The total number of housing starts across the nation in that year amounted to only 5 percent of the 1928 figure. In addition, by 1930 there were 150,000 nonfarm foreclosures, more than double the pre-Depression rate. These foreclosures increased by about 50,000 homes each year until 1933, when half of all home mortgages in the United States were in default. More than a thousand urban residential foreclosures occurred every day.[23] Throughout America, the homeless and unemployed collected in tent cities, hobo jungles, and Hoover Heights.

World War II improved the situation greatly, as 16 million young service men and women shipped out to the European and Pacific theaters. But this improvement was only temporary. After demobilization, the situation worsened again. In Chicago in 1946, for example, the housing crisis was so bad that over 100,000 veterans—mostly unemployed young men—were homeless.[24] In their book *Picture Windows: How the Suburbs Happened*, Rosalyn Baxandall and Elizabeth Ewen eloquently illustrate just how acute the housing crisis had become: "In Atlanta, 2,000 people answered an advertisement for one vacancy. A classified ad in Omaha newspaper read, 'Big Ice Box 7 by 17 feet. Could be fixed up to live in.'"[25]

In 1933, under FDR, the federal government responded to the crisis with a series of experimental moves that would ultimately render old techniques and styles of housing construction obsolete. For the first time, the Federal Housing Administration recommended national space, safety, and construction standards. The Home Owner's Loan Corporation refinanced thousands of loans in immediate danger of default and foreclosure. HOLC also granted low-interest loans that enabled some homeowners to recover lost property. Their major innovation, however, was the introduction of long-term self-amortizing mortgages with uniform payments spread over the life of the debt. This development was to have a remarkable effect on American real estate after 1944, when section 505 of The Serviceman's Readjustment Act (the GI Bill) guaranteed every veteran a fully financed mortgage for any home meeting FHA standards. Then in 1947, the Housing and Rent Act introduced rent controls to keep rental housing affordable. One consequence of this act was to make rental properties much less lucrative for investors, who turned their capital and energies to the construction and sale of privately owned homes.

On the outer rings of cities where land was cheaper, suburban developments sprang up almost overnight, aimed at a new blue-collar mass market.[26] Houses in this new suburbia were pared down to their most essential features, as developers found ways to economize on construction. The new attitude inside the building trade was best expressed by William Jaird Levitt, a law school dropout whose firm constructed the largest and most successful suburbs of postwar America: "We are not builders. We are manufacturers. The only difference between Levitt and Sons and General Motors is that we channel labor and materials to a stationary outdoor assembly line instead of bringing them together in a factory on a model line."[27]

Abraham Levitt had first involved his two sons in construction in 1929. By 1947 Levitt and Sons began building their first of three Levittowns, this one near Hempstead, Long Island. It consisted of row upon row, acre upon acre, of affordable detached houses on 6,000 square foot lots. William Levitt keenly understood and appreciated the inner workings of mass production and obsolescence. A consummate salesman, gambler, and deal maker, William Levitt was also a favorite of Senator Joseph McCarthy. In July 1952 William saw his own face staring back at him from the cover of *Time* magazine, which pegged him as "a cocky, rambunctious hustler." He was the kind of man who could parlay an $80 stake into $2,500 in one short afternoon in a casino by spotting and playing the short odds. But it was Alfred Stuart Levitt who, according to his father, was the real genius of the business. Much less flamboyant than his older brother, Alfred spent ten months with Frank Lloyd Wright in 1936 during construction of the Rehbuhn house in Great Neck, New York, where he learned to streamline the principles of home design and apply them to modern American living spaces, eliminating all that was obsolete in residential architecture.

In 1939 Benjamin Rehbuhn began serving a two-year sentence in Lewistown, Pennsylvania, for pandering, a term that in this case meant using U.S. mail to distribute obscene material published by Falstaff Press, a company owned by Ben and his wife, Anne. (Because she was pregnant at the time, Anne was spared prison but fined $5,000.)[28] The material in question was information about contraception from Margaret Sanger's American Birth Control League, a forerunner of the Planned Parenthood Association. Rehbuhn was well aware of the need to make birth control information freely available. He was youngest of fourteen children of a Viennese Orthodox rabbi who ran a successful—though very crowded—East Side grocery.

Ben was the only child in his family who managed to get an education. He distinguished himself early at City College and then rode scholarships to Columbia's graduate school, where he earned a master's degree in English and American Literature before being drawn to publishing. As a reward to himself for an earlier obscenity stretch in Lewistown, he commissioned Wright to design a Long Island retreat for the Rehbuhn family. He chose Great Neck because it was well away (seventeen miles) from the hustle and bustle of his Manhattan apartment. After that first internment, Ben wanted a home with the ample space and privacy he had dreamed about as a boy. Like many other upwardly mobile Americans of the 1930s, he was drawn to the openness and seclusion of Wright's Usonian homes, and it is not difficult to imagine why.

Alfred Levitt paid $10,000 to observe Wright's construction of the Rehbuhn home. He knew that open-plan Usonian houses would be popular with the same upwardly mobile second-generation families who had bought the two hundred homes in Strathmore-at-Manhasset that the Levitts had built in 1934. But the ideas Alfred took away from Great Neck allowed him—ten years later—to design even simpler houses for the company's veteran and working-class clientele.

During the war, the Levitts prospered with contracts for various Navy housing projects, where they gained more experience in applying mass production techniques to home construction. Steered by Alfred's design vision, William put Walter Gropius's ideas for standardized worker housing into practice. As careful observers of the federal government's Greenbelt Project, the Levitts also learned to use new materials and new tools, including the first power tools (drills, nailers, belt sanders, and circular saws) in home construction.[29] Then, in 1944, while William was away in Oahu with the Seabees, Abraham and Alfred acquired a large tract of land near Hempstead, Long Island. As their project unfolded, they decided to apply Alfred's genius for developing economies of scale to make affordable housing for those who needed it most: inner-city residents who could be lured to the spaciousness of the suburbs.

At first, luring them proved difficult. Hicksville, the train station closest to Island Trees (later renamed Levittown) became such a popular joke among city dwellers that it survives today as a synonym for the sticks. Still, with very little housing to be had in New York City, the Levitts' well-made, inexpensive detached homes found a ready market. The scale of their popularity commanded national attention. A comprehensive study of Levittown in 1952 noted that "despite the fact that Levittown can be viewed . . . as representative of . . . long term decentralization . . . it is assuredly the most spectacular step ever taken in this direction."[30]

In these homes, called Cape Cods, all nonessentials were eliminated. Wright abhorred basements, because they wasted space, time, and money. During visits to Japan he had learned about radiant-heated floors, an idea he brought back to America. Instead of the expense of digging and finishing a basement, Wright's suburban houses were built on a concrete slab that contained coiled copper tubes through which hot water flowed evenly under the floor of the living space. The floors in Alfred's naval housing during the war and, later, his Cape Cods copied the heating system of his teacher's Usonian prototype. Although this heating system was not Alfred's innovation, the publicity that Levitt and Sons derived from the scale and success of the first Levittown on Long Island virtually eliminated basements in all but luxury-class residential architecture in the decade after World War II. In every major metropolis, from Baltimore to San Antonio, large construction companies appeared which adopted Levitt and Son's concrete-slab construction.[31]

If basements became obsolete simply as a matter of economy, the reasons for the appeal of porchless houses was slightly more complex. After World War II, many people came to associate porches with old-fashioned houses whose indoor plumbing, electrical wiring, and other amenities were substandard.[32] But in the postwar years, porches suffered from another unpleasant association. They were one example of what the sociologist Sharon Zukin calls liminal spaces—public areas for meeting, mixing, and transit.[33] For low-income inner-city tenants, these liminal spaces, including front stoops, hallways, parks, sidewalks, squares, bus stops, and train terminals, can be sites of uncomfortable interactions—elevator silences, excessive noise, physical aggression. Especially during a depression or other housing crisis, liminal spaces are always overcrowded. Unemployed adults spend long hours there, as do children, poor relatives, and newlyweds forced to bunk-in with family members because they cannot afford their own apartment.

The appeal of the porchless suburban house was the escape it offered from intense, overcrowded interaction with one's neighbors.[34] Contrary to popular myth, the urban working class was not driven to Levittown by crime; it was lured there by the promise of space and privacy offered by a detached home. These two factors explain not only the disappearance of the porch but the reorientation of suburban homes away from the street and toward the large yard out back. By the 1950s that suburban backyard would be furnished with a barbecue pit and picnic table for dad, a sandbox or gym set for the kids, and maybe a flower bed and hammock for mom. Other Levittown design features emphasizing seclusion included curvilinear streets that made neighboring homes less visible, an absence of sidewalks to discourage foot traffic, and extensive landscaping which, when properly viewed through the ample picture window at the front of all Levittown homes, mimicked a peaceful rural setting.

Barbara M. Kelly, former director of the Long Island Studies Institute, has written about the importance of privacy as an overriding design feature in Alfred Levitt's homes: "The houses of Levittown were designed to be very private. Each was centered on its own plot of 6,000 square feet. The side door—the entrance into the kitchen—[was] on the right wall of the exterior of the house. (The kitchen entry was placed on the right of every house, where it faced the wall of the house next door.) A vestigial eight-foot length of split rail fence separate[d] this 'service' area from the more formal front lawn. Trash was picked up at the street end of the service path in front of each house. There was no communal path, no common service area."[35] Alfred, who was the exact opposite of his gregarious brother, was obsessed with privacy in housing design and expressed his belief that it could be made compatible with the mass production of homes. "Without even a trial balloon you guess in advance just how far you could push and persuade [people] to take either open planning or living in a fish bowl until the landscaping grows to give them privacy."[36]

After 1948, when the housing crisis eased somewhat, the Levitts committed themselves to constructing detached housing for veterans. They could hardly lose. The government guaranteed veterans' mortgages and made them available without any down payment at all. So, just as competition was entering the Levitts' new downscale market, Alfred redesigned the Cape Cod to make it attractive to a middle-class clientele. His newest design featured a ranch-style home available in four basic models. The company then acquired new land parcels and began building middle-class Levittowns. William scheduled regular open-house showings for these ranch homes, announced with full-page ads in all the New York papers.

The Levitts installed the very latest appliances in their model homes: General Electric stoves, Admiral televisions (after 1950), and Bendix washers. The postwar GE refrigerators in Levitt homes included larger freezers, since Clarence Birdseye's process of flash-freezing fish had come into its own during the war and had rendered obsolete the salt-cured and canned foods of yesteryear.[37] With these shiny ultra-modern appliances, the Levitts waged a campaign of psychological obsolescence against veterans and their wives. In 1944 a devoted Milwaukee husband was one of their first victims: "We had a nice little house . . . It was just like hundreds of other houses that were built twenty five years ago. We all liked it fine. One Sunday the wife noticed an ad for a new house. She kept after me to drive up there and we went through it. The old house never looked quite right after we made that inspection. We found we were hopelessly out of date . . . Well, to make a long story short, we got the habit of visiting those new houses. Pretty soon the wife was so dissatisfied that we had to buy one."[38]

By the time they were finished, the Levitts had built three Levittowns, in New York, Pennsylvania, and New Jersey. Eventually, they sold the business for $70 million and the promise that William would refrain from participating in the home construction industry for ten years—a period that proved sufficient to render his expertise itself obsolete. Without Abraham and Alfred to correct his excesses, Bill Levitt gambled away his share of the family fortune and died penniless in a crowded public hospital in New York City. With his last breath, he was still promoting some deal that would restore his fortunes and put him back on top.

ANALOG VERSUS DIGITAL COMPUTING

Any history of technology published before 1973 will list John Mauchly as the inventor of digital computing. Actually, the American whose invention made analog computing obsolete was a little known Iowa State college professor named John V. Atanasoff. Working in isolation and obscurity with a graduate assistant, Charles Berry, in the late 1930s, he invented the Atanasoff-Berry Computer (ABC), the first fully electronic computing machine.[39] The MIT whiz kid Vannevar Bush had often applied the term "analog" to descriptions of the circuits he built in the 1920s, but Atanasoff was among the first to apply the term to a computing device and to distinguish between analog and digital computers.[40] This is an important point because it establishes John Atanasoff at a historic midpoint between Bush's own differential analyzer and ENIAC, the breakthrough device designed and assembled in a frenzy of war-driven effort at the Moore School of Electrical Engineering at the University of Pennsylvania.

The differential analyzer was a device from another age. It used continuous mechanical processes to mimic mathematical equations. (Slide rules are another kind of analog calculator.) A digital calculator, on the other hand, breaks down a computation into a lengthy series of simple addition and subtraction operations that can be performed at great speed because digital machines rely on pulses of energy, not on mechanical moving parts. The story of how the differential analyzer became obsolete during World War II is really the story of the rise of modern computing and the story of how engineering itself changed from a culture "couched in the graphic idiom . . . that emphasized graphical presentation and intuition over abstract rigor."[41]

The machine Bush invented in 1925 could mechanically solve differential equations of motion in several hours, as opposed to days of manual number crunching. A front-page story in the *New York Times* in 1927 announced that this "Thinking Machine Does Higher Mathematics: Solves Equations That Take Humans Months."[42] Solving differential equations quickly was an especially good thing to do because these equations played an important role in ballistics. In order to predict accurately where various shells would land during wartime, the Ballistic Research Laboratory at the Aberdeen Proving Ground in Maryland assembled complex charts called trajectory tables that enabled American gunners to target their guns under a variety of conditions—changes in atmospheric pressure and humidity, wind speed and direction, types of guns and projectiles. The hours of labor involved in the manual calculations required to assemble these tables were staggering and prohibitive.

The BRL assembled its own differential analyzer in Maryland but the machine's design was flawed. It frequently broke down and lost data as it neared the end of a calculation. Frustrated with this constant time-consuming interference, the Ordnance Department contracted with the Moore School of Electrical Engineering to perform its calculations. The Moore School was the proud possessor of a differential analyzer whose features were much more modern and reliable than either the MIT or BRL machines. To support the ballistics project, the Moore School assembled an accomplished team with a variety of backgrounds. This group included John W. Mauchly, the man who was originally credited with the idea of electronic digitization which was at the heart of ENIAC, the hoary old grandfather of all personal computers.

Shortly before Mauchly was recruited for the ballistics project, he became interested in designing a faster calculating machine to make meteorological predictions.[43] He believed that accurate weather prediction would somehow lead to considerable financial success. Although his PhD was in physics, Mauchly had begun academic life by training as an engineer for two years on the advice of his father, a frustrated Carnegie Mellon physicist who understood that good engineers often earned more than good professors. Perhaps because of his father's concerns, Mauchly was preoccupied throughout his life with financial success. And as a young college professor at Ursinus College in Collegeville, Pennsylvania, Mauchly was certainly not earning a lot of money.

Gradually, as his meteorological research progressed, Mauchly realized that a reliable electronic calculator had much wider commercial applications than just predicting the weather. He began to survey the field carefully. At Swarthmore College he learned that vacuum tubes could be used to distinguish between electrical signals separated by only one-millionth of a second. If this ability could be put into the service of a calculator, the machine would be much faster than any available mechanical device.[44]

In 1939 Mauchly took a night course in electronics to learn more about circuitry. He also attended the New York World's Fair, where he saw an IBM cryptographic machine with vacuum tube circuits.[45] At a meeting of the American Mathematical Society in 1940 he witnessed a demonstration of an electromechanical calculator. Then in June 1941 Mauchly took his son on a visit with a young professor at Iowa State College whom he had met the previous December. Although he would deny it later during a legal dispute over patents, Mauchly learned a great deal from John Atanasoff and his graduate assistant, Charles Berry, during his four- or five-day visit to Atanasoff's home and lab. In particular, Mauchly was given hands-on access to a device that was not electromechanical. For all its imperfections, the ABC was unique because it was fully electronic.[46]

Despite his access to Atanasoff, Berry, and the ABC machine itself, Mauchly would later testify that he had spent just "one and one-half hours" or less in the presence of Atanasoff and Berry's computer, and that the machine, sitting over in the shadows, may not have even had its cover off, he said.[47] Unfortunately for Mauchly, John Atanasoff was a packrat who saved all correspondence and records. Twenty-seven years after Mauchly wrote him with observations about the ABC machine and what he had learned during his visit, Atanasoff was able to produce letters in which Mauchly praised the computer for being able "to solve within a few minutes any system of linear equations involving no more than thirty variables." These letters included the observation that the ABC machine could "be adapted to do the job of the Bush differential analyzer more rapidly than the Bush machine does, and it costs a lot less." Mauchly's letter of September 30, 1941, contained this telling passage:

> A number of different ideas have come to me recently . . . some of which are more or less hybrids, combining your methods with other things . . . The question on my mind is this: Is there any objection, from your point of view, to my building some sort of computer which incorporates some of the features of your machine . . . Ultimately, a second question might come up, of course and that is, in the event that your present design were to hold the field against all challengers, and I got the Moore School interested in having something of the sort, would the way be open for us to build an Atanasoff Calculator (à la Bush analyzer) here?

Politely, Atanasoff refused, informing Mauchly that any such agreement might negatively affect his patent application for the ABC. Unknown to Atanasoff, the Iowa State College administration would neglect to pursue his patent. Despite Atanasoff's refusal, Mauchly persisted in adapting the Iowa inventor's ideas. From that point forward, even though Mauchly continued to use Atanasoff as a sounding board and a source for job recommendations, he was completely secretive with him about design improvements he was making in the ABC and about the development of ENIAC. At stake was probably Mauchly's standing at the Moore School, where, before ENIAC, he was regarded as an unsophisticated flake with too many undeveloped ideas on too many subjects. The attitude of his colleagues toward him at this time might also explain why his first proposal for an electronic computer, dated August 1942, was lost soon after it was submitted.[48]

Six months later, when Lieutenant Herman Goldstine, the mathematician charged with BRL's differential analyzer project at Penn, learned of Mauchly's proposal, he asked to see it. "The Use of High

Speed Vacuum Devices for Calculation" had to be entirely recreated from Mauchly's notes in the early weeks of 1943. To everyone's surprise, Goldstine got approval to fund the project by April, and by October 1945, thanks to prodigious circuitry designed by J. Presper Eckert, ENIAC was fully assembled and operational.

In 1945 the ENIAC superseded the newest version of the differential analyzer then being assembled at MIT with funds from Carnegie Corporation and the Rockefeller Foundation. Over the next five years, MIT's project stalled again and again in light of the knowledge that large analog calculations were no longer effective. In 1950 when ENIAC, EDVAC, and UNIVAC had become firmly established, Samuel Caldwell, director of the Center of Analysis at MIT, confessed to Warren Weaver, director of the Natural Science Division of the Rockefeller Foundation, that the analyzer was "essentially obsolete" and that the program to develop it had "become a real burden on MIT."[49] Weaver concurred.

The obsolescence of the differential analyzer not only marked the passing of analog computing but, as Larry Owens notes, it marked the obsolescence of the cultural values of early twentieth-century engineering in which "students honed their problem-solving talents with graphical methods and mechanical methods of various kinds—slide rules, planimeters, a plethora of nomographic charts and instruments for the mechanical integration of areas under curves." Like so many other things, this archaic version of engineering culture did not survive World War II.

Mauchly's reputation as an American inventor also suffered. In a patent suit filed by Honeywell against Sperry Rand, which held the Mauchly patents in 1973, a federal judge ruled that "Eckert and Mauchly did not themselves first invent the automatic electronic digital computer, but instead derived that subject matter from one Dr. John V. Atanasoff."[50] The court formally stripped

Mauchly of his patents. And in 1990, at age eighty-six, John V. Atanasoff received recognition in the form of the National Medal of Technology.

In 1933, one year after Archibald MacLeish's "Obsolete Men" appeared in *Fortune* magazine, Norman Cousins, who was then eighteen, graduated from Columbia University Teacher's College and become one of hundreds of thousands of technologically unemployed men looking for a job in New York City. As a practicing if rarely a church-going Unitarian, Cousins harbored an old-fashioned belief that writers had a social responsibility to provide clear and thoughtful leadership during times of crisis. Articles demonstrating these convictions in the *New York Evening Post* and *Current History* won appreciative notice for the young man.

In 1940 Cousins was offered the position of executive editor for the *Saturday Review of Literature,* and by 1945 the *Saturday Review* had become the *Charlie Rose Show* of its day. It was an informed and influential forum of ideas with a weekly circulation of over 600,000 subscribers. Cousins, now the editor, was a deeply committed humanist who during a long and brilliant career promoted the causes of nuclear disarmament, world federalism, and humane medical practices. On the morning of August 6, 1945, as he sat down to breakfast, Cousins read the screaming headlines in the *New York Times* announcing that the world had changed forever. His reaction to William Lawrence's inside story of the bombing of Hiroshima and the subsequent attack on Nagasaki appeared first as a long editorial essay in *Saturday Review* published on August 18, four days after the Japanese surrender.

"Modern Man Is Obsolete" provoked an enormous response, and the title quickly slipped into the mainstream of public discourse.[51] By the war's end, the leitmotif of mankind's obsolescence had become familiar to every American who read the newspaper or listened to the radio. Cousins's suggestion that the bomb rendered humanity obsolete fell on fertile ground. By October his essay was available in an elongated book form that was reissued many times during the 1940s, even though Americans still overwhelmingly supported Truman's decision to bomb both Hiroshima and Nagasaki.[52] A Gallup poll taken at the end of August 1945 gave an approval rating of 85 percent to the attacks. While this approval declined to 53 percent over the next few months, a significant number of Americans—nearly 25 percent—expressed the wish that America had used many *more* bombs on Japan before it surrendered.[53] A radio comedian made the tasteless joke that Japan was suffering from "a 'tomic ache," and a new drink called the Atomic Cocktail—a mix of luminescent green Pernod and gin—enjoyed a brief surge of gut-wrenching popularity.[54] Clearly, in 1945 America's eventual forgiveness of the Japanese both for Pearl Harbor in 1941 and for the Bataan Death March in 1942 was a long way off.

Still, Americans were concerned that a Pandora's box had been opened and atomic horrors of all kinds had entered the world. Modern science had created the technological unemployment that prompted MacLeish's essay "Obsolete Men" in 1932. But, as Cousins wrote, "The change now impending is in many ways more sweeping than that of the Industrial Revolution itself."[55] By 1945 science threatened to obliterate mankind entirely. From its opening words, Cousins's timely essay speaks directly to these fears:

The beginning of the Atomic Age has brought less hope than fear. It is a primitive fear, the fear of the unknown, the fear of forces man can neither channel nor comprehend. This fear is not new . . . But overnight it has become intensified, magnified . . . It is thus that man stumbles fitfully into a new era of atomic energy for which he is as ill equipped to accept its potential blessings as he is to control its potential dangers.

Where man can find no answer, he will find fear. While the dust was still settling over Hiroshima, he was asking himself questions and finding no answers. The biggest question of these concerns himself. Is war inevitable because it is in the nature of man? If so, how much time has he left—five, ten, twenty years before he employs the means now available to him for the ultimate in self-destruction—extinction . . . It should not be necessary to prove that on August 6, 1945, a new age was born. When on that day a parachute containing a small object floated to earth over Japan, it marked the violent death of one stage of man's history and the beginning of another. Nor should it be necessary to prove that the saturating effect of the new age, permeating every aspect of man's activities, from machines to morals, from physics to philosophy, from politics to poetry; in sum, an effect creating *a blanket of obsolescence not only over the methods and the products of man but over man himself.*

The thrust of Cousins's essay is really an appeal for a new and enforceable world order, one that would prevent global war and nuclear proliferation. Technocracy's old theme of a technology that threatened man and therefore required enlightened management had been reawakened in 1943 when Roosevelt's Republican opponent Wendell Lewis Willkie published a bestseller, *One World.*[56] Gradually, ideas of world government were constellating into what would become, by 1947, the United World Federalist movement. In 1945, the time was exactly right to make such a pitch. Weeks before Germany capitulated, the United Nations

Conference had convened in San Francisco, to enormous international interest and enthusiasm. Around the world, there was renewed determination to create an international forum stronger and more effective than the League of Nations had been between the wars.

Decades before Marshall McLuhan coined the phrase "global village," Cousins clearly understood the planetary implications of the previous years of conflict: "The world has at last become a geographic unit, if we measure geographic units not according to absolute size, but according to access and proximity. All peoples are members of this related group . . . The extent of this relationship need only be measured by the direct access nations have to each other for purposes of war."[57] Cousins's essay included a personal journalistic manifesto that would serve him well for the next forty years. From August 1945, he recognized what his role in the new world would be. "Man is left," he wrote,

> with a crisis in decision. The main test before him involves his will to change rather than his *ability* to change . . . That is why the power of total destruction as potentially represented by modern science must be dramatized and kept in the forefront of public opinion. The full dimensions of the peril must be seen and recognized. Only then will man realize that the first order of business is the question of continued existence. Only then will he be prepared to make the decisions necessary to assure that survival.

What this meant for Norman Cousins was a new determination to use the medium of the *Saturday Review* to promote world peace and world federalism—a cause discounted as starry-eyed idealism by most contemporary journalists.[58] Few readers will now remember Cousins's famous account of his trip to Hiroshima and how he appealed to *Saturday Review*'s readership to sponsor twenty-four "Hiroshima maidens" for the treatment of radiation sickness in the United States. (True to his convictions, Cousins and his wife later adopted one of these young women.) Because of the overwhelmingly favorable response of his readership, Cousins was also able to pay for the medical care of about four hundred children orphaned by the Hiroshima bomb.

Articles describing these efforts helped present the human face of the Japanese people to postwar America. They also fulfilled Cousins's personal program of keeping the horrors of atomic war front and center while attempting to provide an alternative perspective on the arms race through his writings and through organizations like the United World Federalists and SANE, the National Committee for a Sane Nuclear Policy, which chose its nonacronymic name because of its sharp contrast with MAD, mutually assured destruction, the ugly reality underlying the cold war arms race.

Less well known than "Modern Man Is Obsolete" is Cousins's piece concerning the nuclear tests at Bikini Atoll in the Marshall Islands.[59] In 1946, what was then called Operation Crossroads tested two 23-kiloton atmospheric nuclear weapons. Efforts to curtail military spending had begun just hours after the Japanese surrender, with the cancellation of hundreds of millions of dollars in wartime manufacturing contracts. But the government was anxious to economize further. Because atomic bombs could be delivered aerially, politicians and military strategists began to discuss the possibility that the large imperial naval forces of the prewar era were obsolete. The navy, however, was still an extremely powerful entity. To end any "loose talk to the effect that the fleet [was] obsolete in the face of this new weapon," Lewis Strauss, aide to then secretary of the navy James Forrestal, recommended that the navy "test the ability of ships to withstand the forces generated by the atomic bomb."[60] In essence, the power and decisiveness of atomic weapons had generated a turf war among the service branches of the military.

The army, which then included the army air force, clearly had a central role in the delivery of the atomic devices that had ended the war. This left the navy in a new and uncomfortable position as a secondary branch of the military whose role and funding would be progressively curtailed. Secretary Forrestal was worried that what the Japanese had been unable to do to the American fleet at Pearl Harbor would be accomplished by cost-cutting and political wrangling in the years following the war. In order to prevent the obsolescence of the navy, and in order to play for a bigger role in the delivery of atomic weapons, three tests of nuclear devices were scheduled for Bikini Atoll in the Marshall Islands. The tests were called Able, Baker, and Charlie. Of these, only the first two were completed.

Cousins's role in the tests was that of an eyewitness, one of thousands who descended on the Marshall Islands to document the blasts. (In a strange coincidence, John V. Atanasoff was also there, in charge of the navy's acoustic measurements.) Cousins had a very real concern that after more bombs were detonated, the public would become habituated to their existence, and atomic destruction would "dissolve into a pattern." His fears were confirmed when a swimsuit made of tiny strips of fabric suddenly achieved worldwide popularity. The French designer, Louis Reard, wittily renamed an existing swimsuit by Jacques Heim the "Bikini" because it reminded him of the insignificant strips of land in the Marshall archipelago that were vaporized by hydrogen bombs. More than anything, Cousins wanted to prevent such normalization and public acceptance of atomic blasts, a process he called "the standardization of catastrophe."

The second Bikini test was the most successful. In the Able test only two of the seventy-three guinea-pig ships were destroyed because the bomb missed its target. In test Baker, twenty-three days later, sixteen of the ships survived detonation but all of them were thoroughly "hot"—too radioactive for living things. Both explosions produced an atomic fog that made it impossible to witness the immediate effect on the target ships. Officers and scientists had been issued protective goggles, so it was only enlisted men who saw the deep copper color of the atomic blast before the five-foot waves came out of the mist and jarred everyone aboard the observation vessels. After the second test, Cousins warned: "This is the supreme weapon against human life. The millions of degrees of heat generated by the protesting atom and the overwhelming blast are only two aspects of the danger. Even more menacing is the assault upon human tissue, upsetting the rate of growth and condemning many thousands outside the area of total destruction . . . The next war, if it comes will be fought not with two atomic bombs but with hundreds, perhaps thousands, of atomic bombs."[61]

Much more famous than Cousins's account of the two Operation Crossroads tests on Bikini Atoll was *No Place To Hide*, a book written by David Bradley, a physician and journalist who had been in charge of the Radiological Safety Unit at Bikini. Published in 1948, *No Place To Hide* became a book-of-the-month selection and remained on the bestseller list for ten straight weeks. As a first-hand account of Operation Crossroads, Bradley's narrative was excerpted in *Atlantic Monthly* prior to publication and later condensed by *Reader's Digest*. If the Bikini tests had rendered

Americans complacent about the dangers of atomic war, *No Place To Hide* woke them up. Cousins wrote of it admiringly: "At a time when the world has virtually accepted the inevitability of another atomic war . . . Bradley states flatly, as other atomic scientists have done, that there is no defense against atomic attack."[62]

Fortunately for us all, Cousins's prediction proved false. Mankind was imperiled but not yet obsolete. Despite the increasing complexity of our world, modern men and women still exist, as does the U.S. navy. World federalism, however, did not last much beyond the 1940s. In 1949 Cord Meyer, who had been elected first president of the United World Federalists when it was founded in 1947, withdrew from active membership. Later that year he joined the CIA.[63] Cousins, who had also served as president of the organization, wrote with considerable sadness about the world federalists' inability to overcome the resistance of entrenched powers to the idea of subordinating themselves to a democratic, international body. Throughout the remainder of his life he campaigned vigorously against nuclear proliferation and all forms of war.

It is deeply ironic that a movement purporting to be a revolt against the conformist sartorial codes of mass society wound up providing such a powerful fuel for nothing other than obsolescence.

THOMAS FRANK, *THE CONQUEST OF COOL* (1997)

To end one chapter with Norman Cousins and begin the next with Brooks Stevens requires ethical yoga. Like Cousins, Stevens left school and started to work in 1933, one year after Bernard London issued *Ending the Depression through Planned Obsolescence.* Unlike Cousins, however, Stevens was the unapologetic child of corporate culture who never turned down a paying assignment over small matters like tastelessness. If he is remembered at all today, it is probably for his 1958 redesign of the original 1936 Oscar Mayer Wienermobile. Stevens's definitive contribution to this cultural icon was to "put the wiener in the bun."[1]

Stevens's father was a prominent VP at an electric motor control manufacturer in Milwaukee. Privileged and smart, Stevens attended Cornell's architecture school and then returned to Milwaukee to practice industrial design. His family's influential contacts served him well, but he was also genuinely talented. His gleaming Edmilton Petipoint clothes iron of 1941 is truly beautiful and still prized by collectors. The design of his clothes dryers and refrigerators paid homage to the early work of Norman Bel Geddes, Raymond Loewy, and Walter Dorwin Teague. Even products that were never manufactured and have only a paper existence, such as his rendering of a unique front-loading toaster, the remarkable Toastalator of 1942, reveal a special flair.[2] In recognition of his talents, Stevens was one of only ten people selected to become charter members for life in the Society of Industrial Designers when it was founded in 1944.[3]

In the late 1940s Stevens's firm designed the Jeep Station Wagon, the distinctive Olympian Hiawatha train, and the Harley-Davidson FL Hydra-Glide or "Panhead." Then, in 1951, Alfa Romeo hired him as a consultant for its 1800 Sprint Series— a coup that fed his ambition to become an internationally recognized automobile designer. At considerable personal expense, he designed a distinctive sports car built on a Cadillac chassis by a coachbuilder favored by Rolls Royce and Mercedes-Benz. From Stevens's design, Herman Spohn of Ravensburg built Die Valkyrie, which debuted successfully at the Paris Auto Show in 1954.

On the occasion of his eightieth birthday in 1991, a *Chicago Tribune* retrospective began with the words "Brooks Stevens is hardly a household name."[4] But that was not the case in the 1950s, when he was recognized as America's controversial "crown prince of obsolescence."[5] Stevens claimed—publicly and often—that it was he who actually invented the phrase "planned obsolescence," and he was certainly the term's most vocal champion. Due to his efforts at self-promotion, many people today still believe the phrase was born in the era of tailfins and Sputniks. But in fact it was used, if not coined, by Bernard London back in 1932, and by the time of a 1936 article on "product durability" in *Printers'*

Ink, the phrase was in common usage among marketing people.[6] Stevens probably first encountered "planned obsolescence" through discussions with other designers. He first used the phrase in print around 1952, in a self-published brochure.[7]

Planned obsolescence, for Stevens, was simply psychological obsolescence, not product death-dating. It grew out of "the desire to own something a little newer, a little better, a little sooner than is necessary."[8] Stevens was aware of earlier work on the subject, especially *Selling Mrs. Consumer,* and many of his arguments favoring planned obsolescence repeat those of its author, Christine Fredericks. In particular, he was fond of claiming that planned obsolescence stimulates the economy, and that quickly dated products are not wasted because they are resold and redistributed.[9] After his Paris Auto Show success, Stevens began to step up his speaking engagements, interviews, and position pieces on planned obsolescence.[10] He enjoyed his new reputation as the bad boy of industrial design, and he used this carefully constructed image to garner more publicity for himself and his work. His firm was patronized by top-level executives, who felt he understood their position very well.

And what exactly was the corporate position on planned obsolescence? In a 1958 interview with Karl Prentiss in *True Magazine,* at a time when America's wastefulness had blossomed into a national controversy, this was Stevens's answer: "Our whole economy is based on planned obsolescence and everybody who can read without moving his lips should know it by now. We make good products, we induce people to buy them, and then next year we deliberately introduce something that will make those products old fashioned, out of date, obsolete. We do that for the soundest reason: to make money."[11]

TAILFINS, EDSELS, AND AUTOMOBILE OBSOLESCENCE

By the end of the Korean War in 1953, the country was riding a wave of prosperity, causing Eisenhower to emphasize the business-friendly atmosphere of his administration and in particular the importance of the automobile industry to the American economy. Making good on his campaign promise to end "the obsolescence of the nation's highways [that] presents an appalling problem of waste, danger and death," Eisenhower started construction of a network of modern roads "as necessary to defense as it is to our national economy and personal safety."[12] Better roads and bigger cars were a natural match, and nobody understood that better than Cadillac.

Inspired by the P-38 Lightning Fighter, tailfins were first introduced on the 1948 Cadillac by Frank Hershey, one of Harley Earl's designers at General Motors. Earl himself did not like fins at first and ordered them removed from the 1949 Cadillac, only to reverse himself suddenly when consumer studies indicated the public loved them.[13] Like their spiritual predecessors—the curled toes on medieval aristocrats' shoes—tailfins grew progressively more exaggerated in the models that followed. In the fall of 1952 GM debuted its 1953 Cadillac Eldorado in a stylish limited edition with a wraparound windshield, a distinctive bumper (including busty "Dagmars"), and enlarged tailfins. During his Inauguration Day Parade in Washington D.C., Eisenhower himself was driven in a 1953 Alpine White Eldorado convertible. With tailfins now the fashion, other automakers quickly followed suit. By 1955 even Brooks Stevens had incorporated tailfins into his custom-designed Gaylord Grand Prix.

Tailfins reached their peak of excess in the 1959 Cadillacs, but by that time it was clear that GM's designs were out of step with

America's changing sensibilities. And there were signs that Earl himself was growing increasingly out of touch with contemporary automotive trends. In the fall of 1956, the "forward look" of Virgil Exner's 1957 Chrysler models introduced a fresh new direction into automobile styling at a moment when Harley Earl was literally vacationing. David Holls, a GM designer of the period, recalled his design colleagues' reactions to their first glimpse of Exner's Chrysler Imperials when the new cars appeared in showrooms: "These cars had absolutely razor-thin roofs and wedge-shaped bodies, and it was just absolutely unbelievable. We all went back [to GM] and said 'My God, they [Chrysler] blew us out of the tub.' And here [at GM] were these ugly, heavy, old-fashioned looking things."[14]

With Earl on vacation, his young designers scrapped their current drawings and began working on new ones that reflected Exner's cars. When Earl returned, he accepted the minor revolution in his shop, realizing that he had indeed lost touch with his customers. He took a less active role soon afterward, waiting out his tenure at GM until his mandatory retirement in December 1958. The 1959 Cadillacs, designed before he retired, were the last GM cars with tailfins.

Meanwhile, at Ford, designer George Walker was trying to avert a disaster comparable to the famous "pregnant Buick" episode of 1929 when GM engineers so badly mutilated one of Earl's designs that the car died in the marketplace. Scathingly, Walker criticized one of Ford's mid-1950s models. In an extremely competitive market, Ford's top executive, Henry Ford II, surprised everyone by listening to him and pulling the car from the assembly line. Walker's doggedness earned him a vice-presidency, an $11.5 million styling center, and considerable in-house power at Ford. But this triumph merely postponed disaster. In July 1956 the stock market took a sudden downturn, and Ford, ignoring the signs of impending recession, unveiled the Edsel in the early fall of 1957, before the company's own end-of-season sales were over. Anxious consumers were able to compare Edsel's high ticket price to the reduced year-end prices of the 1957 models, rather than to the higher prices of its 1958 competitors, and they turned away in droves. The Edsel, which actually was a fine automobile for its time, died in the showroom.[15]

Other external factors entered the consumer equation as well. Just after the Edsel's appearance, the Soviets launched Sputnik I. More than any other mid-century event, this was a tipping point, after which America began a period of introspection that challenged the consumer ethic of waste. Prior to that time, Americans for the most part had been happy to participate in consumerism because they had a vested interest in believing what they had been told: that the wastefulness of planned obsolescence fueled a competitive research economy which guaranteed them a position on the cutting edge of technology. American planes and bombs, science and technology, kept the nation and the world safe, and at the same time provided a constantly increasing level of comfort to more and more Americans.

The launch of Sputnik in October 1957, at a moment of economic recession, was a propaganda coup of the highest order for the Soviets. It challenged head-on two of the most basic premises of American ideology: technological superiority and the economic prosperity it supposedly fostered. As Marshall McLuhan would later observe, "The first sputnik . . . was a witty taunting of the capitalist world by means of a new kind of technological image or icon."[16] The fact that the United States' own Vanguard satellite exploded on the launch pad two months later only deepened

America's moment of self-doubt and readied the country for a period of genuine self-criticism.

While the early retirement of the Edsel was not itself an example of planned obsolescence (indeed, it was very much unplanned), this rejection marked a turning point in the American consumer's previously uncritical acceptance of the ethic of waste. In the fall of 1957, following Sputnik, Americans turned against Detroit's excessive creations. Purchasing remained steady in other consumer areas, but in automobiles—the leading technology of the day—sales fell significantly.

GEORGE NELSON AND VANCE PACKARD

New voices began to challenge the wisdom of waste and the practice of discarding still usable products. This change can be seen in a comparison of the work of two writers, George Nelson and Vance Packard. Now most famous for his distinctive Herman Miller furniture designs, George Nelson published a thought-provoking piece on obsolescence in *Industrial Design* in December 1956.[17] Like his designs, Nelson's writings were much more developed and considered than Stevens's sensational magazine pieces or brochures. He was by turns an associate editor, managing editor, contributing editor, and editor-in-chief of several major design and business publications, including *Architectural Forum, Fortune Magazine,* and *Design Journal.* While clearly in favor of planned obsolescence, Nelson's writing lacked Stevens's cynical opportunism. Also, he did not pander to the corporate elite.

Nelson was equally comfortable in American and European design circles, and he held highly developed and informed opinions about planned obsolescence. These would come to fruition decades later in Milan, through an anti-establishment avant-garde design movement called Memphis, led by his young friend and protégé Ettore Sottsass Jr.[18] In the 1980s Memphis and other anti-modern design groups would give us "chairs that could not be sat on [and] bookcases that could not hold books."[19] Sottsass's witty antifurniture was never meant to satisfy a utilitarian purpose, of course, but was rather designed to accelerate obsolescence to the point of absurdity by making it take place *prior* to the purchase of these articles of furniture.

The origins of Sottsass's aesthetic can be found in George Nelson's 1956 observations about the need to accelerate obsolescence:

> Actually, for all the talk about it we have precious little to show that can be described as planned obsolescence. The traditional city has been obsoleted, in a very real way, by the automobile. But what has happened just happened; there was no planning. This is where real waste occurs and this is what we are beginning to realize: obsolescence as a process is wealth-producing, not wasteful. It leads to constant renewal of the industrial establishment at higher and higher levels, and it provides a way of getting a maximum of goods to a maximum number of people. We have learned how to handle obsolescence as a prodigious tool for social betterment in those areas where we have both knowledge and control. The waste occurs where obsolescence is both too slow and too haphazard, where adequate information and adequate controls and systematic elimination are lacking. We do not need fresh technologies to show us how to upgrade housing—but we do need a continuing method for getting rid of the production we have outmoded. The same holds for cities. What we need is more obsolescence, not less.[20]

At the other end of the spectrum from the privileged and urbane George Nelson was the magazine journalist Vance Packard. Between the ages of twenty-eight and forty-two, Packard honed his craft as a popular journalist for *American Magazine,* a mass circulation middle-brow monthly of the same stamp as the *Saturday Evening Post, Look,* and *Colliers.* Shortly after the stock market downturn in 1956, *American Magazine* folded, leaving Packard scrambling for a way to support his family. He found another position, this time at *Colliers,* but by December *Colliers* itself folded.

Although writing jobs were opening up in television, Packard was convinced he could never write for TV. Instead, he went on unemployment insurance early in 1957, determined to make the transition from magazine to books. He was lucky to have a sympathetic contact at a book publisher. Eleanor Rawson, a former *American Magazine* colleague, had secured an editorial position at David McKay. When the company published Packard's manuscript in the fall of 1957, it had no idea what a runaway success the book would be. *The Hidden Persuaders* examined the claim that scientific advertising (then called motivational research) was essential to America's economic health. The book's claim that American consumers' "buying habits were directly controlled through subliminal techniques" formed the basis of a perspective on consumerism that would come to be called manipulationism.

Though generally dismissed by mainstream sociology today, manipulationism, with its Orwellian connotations, struck a chord with anxious consumers as the economy slowed and the Cold War entered its second decade.[21] Packard's experience writing for magazines allowed him to produce an entertaining and highly readable description of why advertisers needed to create planned or psychological obsolescence: "One big and intimidating obstacle . . . was the fact that most Americans already possessed perfectly usable stoves, cars, TV sets, clothes, etc. Waiting for these products to wear out or become physically obsolete before urging replacements upon the owner was intolerable. More and more, ad men began talking of the desirability of creating 'psychological obsolescence.'"[22]

With the enormous success of *The Hidden Persuaders,* Packard found himself launched on a kind of career that was barely recognized in his own time. A strange combination of social critic, pop psychologist, and quasi–public intellectual, Packard hastily constructed books that would prefigure popular works by Rachel Carson, Betty Friedan, John Kenneth Galbraith, Jules Henry, Christopher Lasch, Marshall McLuhan, and Ralph Nader. Packard was the first writer to catch this wave. In just three years, he produced three nonfiction bestsellers in a row, a feat no other American writer has equaled, before or since. *The Status Seekers* (1959) was a groundbreaking examination of America's social and organizational dynamics, and *The Waste Makers* (1960) was a highly critical book-length study of planned obsolescence in contemporary American culture.

At the appearance of *The Hidden Persuaders,* as America fell into recession, the debate over planned obsolescence exploded into a national controversy. In 1958 similar criticisms appeared in Galbraith's *Affluent Society.* By 1959 discussions of planned obsolescence in the conservative pages of the *Harvard Business Review* created a surge of renewed interest in Packard's first book, which had contained numerous observations about planned obsolescence. With the topic now achieving national prominence, Packard wanted to return to it in a book-length study focused specifically on waste.

Packard first conceived of *The Waste Makers* during a chance encounter with William Zabel, a Princeton student who had written a lengthy undergraduate paper on the topic of planned obsolescence.[23] Packard hired the young man as his research assistant and gave him full credit for his work when the book was published. Benefiting from Zabel's painstaking research, Packard was much more systematic in his approach to obsolescence this time around. Where *Hidden Persuaders* had identified planned obsolescence simply as another name for psychological obsolescence, *The Waste Makers* made much finer distinctions:

> The phrase "planned obsolescence" has different meanings to different people. Thus many people are not necessarily defending deliberately shoddy construction when they utter strong defenses of obsolescence in business . . . we should refine the situation by distinguishing three different ways that products can be made obsolescent . . .
>
> Obsolescence of function: In this situation an existing product becomes outmoded when a product is introduced that performs the function better.
>
> Obsolescence of quality: Here, when it is planned, a product breaks down or wears out at a given time, usually not too distant.
>
> Obsolescence of desirability: In this situation a product that is still sound in terms of quality or performance becomes "worn out" in our minds because a styling or other change makes it seem less desirable.[24]

Although contemporary sociologists often criticized Packard for his lack of scholarly rigor, *The Waste Makers* helped Americans turn a corner in their examination of American business practices. It explained complex ideas, like those embedded in the term "planned obsolescence" itself. But also, where the *Hidden Persuaders* had implicated only advertisers, manufacturers, and marketers, *The Waste Makers* placed responsibility for waste on the consuming public itself. This display of integrity was not lost on the baby boomer generation. Like many of her peers, Barbara Ehrenreich read Packard's books as an undergraduate and later described him as "one of the few dissenters from the dogma of American classlessness."[25] Packard saw the hypermaterialism of American consumer culture as a compensation offered by successful capitalists to the middle and lower classes, whose postindustrial jobs were becoming increasingly meaningless. Motivated by greed born of advertising, Americans conspired with market researchers to transform themselves into "voracious, wasteful, compulsive consumers."[26]

Response to the book from the business community was immediate and hostile—a fact that delighted both Packard and his publisher. Because a short period of seventeen months separated the release of *The Status Seekers* and *The Waste Makers,* Packard had worried that both he and the book might suffer from overexposure.[27] But in 1960 and 1961 Justus George Frederick's old trade journal, *Printers' Ink,* targeted *The Waste Makers*—and Packard personally—in three separate issues, including a special edition whose enticing cover copy read "Is 'The Waste Makers' a hoax? Why did Vance Packard write it? Why did David McKay Co. publish it? *Printers' Ink* and seven outstanding ad men—Cummings, Frost, Zern, Kerr, Weir, Mithun and Cox—explore these questions to try to find reasons for such a deliberate attack on advertising and marketing. The special report begins on page 20."[28]

Although the book was number one on the *New York Times* bestseller list for only six weeks, the business community's hysterical criticisms of *The Waste Makers* attracted enough attention to hold a spot on the list for six months. In October, *Printers' Ink* ran a second feature entitled "Packard Hoodwinks Most Reviewers."[29] Their most vicious attack, however, came six months after the book's release. The article "Has Packard Flipped?" confirmed *The Waste Makers* importance as an effective if mostly solitary attack

on America's growing "throwaway culture."[30] Because of the attention the book excited within America's business community, foreign demand for *The Waste Makers* grew until it was eventually
translated into more than a dozen languages. But, as Daniel
Horowitz notes in *Vance Packard and American Social Criticism,*
"no newspaper reviewers seconded the diatribes from the business community."[31]

Packard's work had far-reaching effects during the 1960s. According to one of the founders of Students for a Democratic Society, Todd Gitlin, Packard's books were among those by "popular
social critics . . . lying on the coffee tables of many a curious adolescent" during the 1950s.[32] Small wonder then that Packard's criticisms of obsolescence were included in the SDS's appeal for social
activism on the part of young Americans. Their 1962 *Port Huron
Statement* read: "The tendency to over-production . . . of surplus
commodities encourages 'market research' techniques to deliberately create pseudo-needs in consumers . . . and introduces wasteful 'planned obsolescence' as a permanent feature of business
strategy. While real social needs accumulate as rapidly as profits, it
becomes evident that Money, instead of dignity of character, remains a pivotal American value and Profitability, instead of social
use, a pivotal standard in determining priorities of resource allo-
cation."[33]

A minor but telling effect of Packard's book was to reinforce
the determination of GM executives to change their vocabulary.
After Packard's first book in 1957, planned obsolescence was
increasingly referred to as "dynamic obsolescence." Thanks to
Packard in the late 1950s, planned obsolescence came to have the
same negative connotations that the earlier term, adulteration,
had had in the previous century.[34]

DEATH-DATING

Shortly after Sony began shipping pocket radios to the United States in 1958, American radio manufacturers went public in the pages of *Design News* with their strategy of death-dating. In the "Design Views" section of this trade magazine for design engineers, editorial director E. S. Stafford wrote: "It is of remarkable interest to learn from a highly placed engineer in a prominent portable radio manufacturing company that his product is designed to last not more than three years."[35]

The fact that durable goods were designed to fail after three years should have surprised no one. During World War II, the Commerce Department had set the minimum requirement for a durable good at three years.[36] But a decade later Stafford wondered if "*purposeful* design for product failure [is] unethical." He suggested it was not, offering the radio engineer's arguments defending his unnamed company's policy of deliberately death-dating their radios: "First, if portable radios characteristically lasted ten years, the market might be saturated long before repeat sales could support continued volume manufacturing, thus forcing the manufacturer into other lines; second, the user would be denied benefits of accelerated progress if long life is a product characteristic."[37]

Although these arguments repeated Sheldon and Arens's main points, Stafford seemed less willing than Brooks Stevens to draw fire. Like GM executives, he clearly avoided the negative connotations of the term "planned obsolescence." Many of the awkward phrases of Stafford's *Design News* piece result from this decision. Maneuvering around the phrase with some difficulty, Stafford inelegantly suggested that

> Planned existence-spans of products may well become one of the greatest economic boosts to the American economy since the origination of time-payments. Such a philosophy demands a new look at old engineering ethics. Respected engineers have long sought to build the best, or the lightest, or the fastest, or at the lowest cost—but few have been called upon to provide all of this—with a predetermined life span. It is very possible that a new factor is entering the economic scene . . . This new factor is Time, in a new costume, requiring new technologies, new concepts—perhaps new ethics. Is this concept bad? We don't think so.

Later on the same page, Stafford offered the following Darwinian argument to counter the anticipated moral objections of consumers: "The consumer might well object to the fact that in a ten-year period he has had to buy three portable radios rather than one. Although he would admit his last radio was more attractive, lower priced and performed much better than the first . . . In this instance . . . 'forced feeding' has contributed to progress."[38]

Despite its forced phrasing, the provocative rhetoric of "Product Death Dates—A Desirable Concept?" had a positive effect. As one reader pointed out, it was "quite stimulating" to air this tacit but important issue. Luckily, this controversy provoked reactions to product obsolescence by working engineers, so it provided a useful cross-section of contemporary opinion among the most literate of these men. Presumably, it also sold magazines. Enough letters from product designers flooded in to *Design News* for this issue to preoccupy the "Design Views" section of three more issues.

Much, but not all, of this reaction was strongly unfavorable and revealed deep antagonism between design engineers and their corporate managers. Harold Chambers from Remington Rand wrote that

> the ready acquiescence evidenced by your editorial . . . is highly regrettable. I greatly doubt that any one of us would wish to apply this "principle" of planned short-term failure to his own purchase of home, auto, piano, and other durable goods involving considerable expense. Why, then, support pressing this principle on someone else? Who is to decide just how short "short term" is? . . . Is not the problem of market saturation a management rather than an engineering problem? . . . Ethics, honesty, truth and other intangible traits are not changeable by management directive!!! Artificial stimulation based on such deliberate dishonest design objectives is certainly a compromise with ethics.[39]

One engineer's letter doubted that the radio firm which favored limiting the life span of its portables really had "the ability to build a product as good as its competitors." He suggested they had developed the idea of a life span for their inferior product as a "naive rationalization for this lack of ability." Another letter expressed shock that an American business had fallen "to such low standards of business ethics." Another said, "The only practical way to reduce the span of a mechanism is to reduce . . . safety." Still another called death-dating "dishonest, immoral and self-destructive both economically and politically [and] a crime against the natural law of God in that we would waste that which He has given us."

Only one letter actually favored planned obsolescence. G. J. Alaback of Whirlpool, a senior manager, presented the manufacturers' case in lucid, reasonable terms. He avoided the suggestion made by Stafford and other letter writers that the purpose of death-dating is to encourage "early repeat business" and instead claimed that mechanical products are essentially not durable, so it is in the consumer's best interest to manage the life spans of each of their components:

> Consumer durable goods need to be designed for some finite life in the consumer's interest. Without a design-life goal, parts of the product might last far longer than others and incur a needless cost liability in the process. Setting the actual life objective is certainly a policy decision faced by a company's top management including of course, engineering. It will undoubtedly vary from one project to another and perhaps would be reviewed and changed from time to time . . . as conditions change. In my experience, a ten or fifteen year design-life is much more common than the one mentioned.

Despite the overwhelmingly unfavorable responses from the professional engineers who made up the majority of its readership, *Design News* itself came down on the management side, favoring planned obsolescence. The January 19, 1959, issue contained a further editorial by Ernest Cunningham, its executive editor and Stafford's boss. Cunningham wrote antagonistically, suggesting that many of his readers' letters "indicate a majority of products are designed for a nearly infinite service life. The truth of the matter is that many engineers do not know the life expectancy of their own product. The principle of Product Death-Dates applies to nearly all design work whether engineers are conscious of it or not. Very often those disclaiming its desirability are operating within its basic principles."[40]

The editorial goes on to commit itself unequivocally to a manufacturing ethic of disposability, providing a benchmark for just how much the idea of durability had changed since the days of Thomas Edison and Henry Ford:

The product with the longest life period is not automatically the
most economical. Value is a product of time and utility. Diminishing returns is an important part of the economic law of supply
and demand and applies to product death dates. Is a product that
has served a short, useful lifetime at a satisfactory cost necessarily
wasteful? I think not . . . There is not a product on the market today that could not be improved by using . . . more expensive materials. Every design is a compromise. Is it wrong, therefore, for
designers to be cognizant of the result and to make the compro-
mises accordingly? Certainly not.

It is impossible to know exactly why Cunningham published
an editorial that really amounts to a position paper on the throwaway ethic. But one *can* speculate. Judging from their published
letters, the *Design News* readership was clearly opposed to death-
dating. But the magazine's corporate sponsors—Whirlpool, for
example—held a much different view and could easily have ex-
erted the usual advertising pressures. Officially, then, the executive editor would have been compelled to uphold the advertiser's
opinion in order to preserve funding for his magazine. But to appease its professional readership, *Design News* needed to return to
the controversy at a later date, using a source from outside the
magazine to express an opinion with more appeal to readers.

In fact the journal did just that, featuring a guest editorial in a
February 1959 issue by Jack Waldheim, senior partner at a design
and engineering firm in Milwaukee. Although Brooks Stevens had
understood planned obsolescence to mean psychological obsolescence (making consumer goods appear dated through the use of
design), Waldheim's piece, by a committed former teacher of industrial design, explicitly concerned death-dating. It was the magazine's final word on the subject of planned obsolescence, a term
that appeared in the *Design News* debate for the first and only

time in Waldheim's essay. And of course it is significant that *Design News* chose not to run a piece in favor of planned obsolescence by Milwaukee's crown prince of obsolescence, Stevens himself. By 1959 planned obsolescence had become a very unpopular business strategy.

"Believe me," Jack Waldheim wrote with considerable conviction,

> your life may be endangered by the spreading infection of planned obsolescence. Planned obsolescence is the deliberate attempt to have something break down or become outdated long before it has lost its usefulness—its utility—or its value! . . . Its danger to me personally is that such sophism on the part of the spokesmen for our profession can kill with distrust the public respect for our skill. Its danger to the customer is that it cheats him out of his hard earned money though he may not realize it in the beginning. If we give the customer what we make him believe he wants we are placing ourselves in the position of expertly skilled con-men . . . planned obsolescence . . . is truly obsolete planning.[41]

Under this abject condemnation of planned obsolescence, *Design News* printed the following disclaimer: "The views expressed are those of Mr. Waldheim and do not necessarily reflect the opinions of the Design News editors." With their advertisers appeased and their circulation now secure, the magazine fell forever silent on the subject of death-dating.

Not so Brooks Stevens. But it was getting harder and harder for him to find a venue that would run an unmediated piece favoring planned obsolescence. Outside the business world, thanks to Packard, planned obsolescence was generally condemned, having become something of a catchphrase for all that was wrong with

America. Never at the forefront of academic discussions of industrial design, Stevens had placed his occasional pieces on planned obsolescence in more popular publications like *True* magazine and Milwaukee's own dailies. The last of these appeared in *The Rotarian* in February 1960.

While lacking the cachet of magazines that published George Nelson and Vance Packard—magazines such as *Industrial Design, Architectural Forum, Fortune, American Magazine, Colliers*—*The Rotarian* nonetheless reached a large number of Americans and provided Stevens with the exposure he craved. Oriented toward small-town businessmen, *The Rotarian* championed the values of citizenship and fair play, and does so to this day. When the magazine ran Stevens's feature on planned obsolescence in February 1960, it gave him free rein to express his views, but balanced his opinion by running his piece as an introduction to a slightly longer essay by Walter Dorwin Teague, an industrial designer whose considerable international reputation outshone Stevens's.

The cover copy "Planned Obsolescence—Is It Fair? Yes! Says Brooks Stevens. No! Says Walter Dorwin Teague," did more to simultaneously popularize and condemn the phrase "planned obsolescence" than any article before or since.[42] And by appearing in a widely circulated magazine that occupied a position of prominence in waiting rooms across the nation, the phrase served as brilliant promotional material for *The Waste Makers*, which had been on the stands for several months. The feature contains very little that is new from Brooks Stevens. Much more important in terms of the social reception of planned obsolescence is Teague's longer and unequivocal condemnation: "When design is prostituted in this way," he wrote, "its own logic vanishes and queer results appear."

One of America's first industrial designers, Teague had won

distinction as the creator of Kodak's truly beautiful Art Deco Brownie (the "Beau" Brownie of 1929) and of Texaco's still recognizable red star corporate logo. He also designed the original interiors for Boeing aircraft and was a cofounder of the American Association of Industrial Designers in 1944. Although not best known for his written work, Teague cared passionately about the subject of planned obsolescence, which he saw as a threat to the integrity of his profession: "This practice of making previous models look outmoded when the new models have no better service to offer is known as 'planned obsolescence' or 'artificial obsolescence,' the latter is the more accurate term but still not as accurate as just plain 'gypping.'"

Teague's article went on to distinguish between Stevens's brand of obsolescence and a more natural variety: "America's phenomenal progress in economic and material welfare has been based on honest, legitimate obsolescence, which is a negative way of saying we advance by making more things more serviceable, less costly. As wise old Henry Ford used to say, 'We aim to make better things for less money.'"

Teague's *Rotarian* article placed a great deal of blame for planned obsolescence squarely on the shoulders of automakers and their designers. Without specifically mentioning Harley Earl, Teague referred to a "hurtful instance of public reaction [to design excesses] . . . when Americans refrained from buying new automobiles. The sales curve took such a bender that it brought on a minor depression . . . You can try every way you like to explain away this buyers' strike, but the facts were plain to . . . any unprejudiced eyes: people in large numbers simply didn't like the typical American cars they were offered . . . If you inquired around you got a variety of answers all adding up to similar conclusions: 'They're hideous' . . . 'I don't like those silly fins.'"

Writing in 1960, Teague was clearly delighted that the tailfin era and the reign of Harley Earl had now ended. Older than Stevens, Teague's vision extended further into the future, and his larger historical perspective illuminated his comments about the effect of GM's uniquely American invention, the annual style change, on the profession of industrial design: "It is of course, impractical, in any major products such as automobiles or high ticket household appliances, to produce really new, retooled models having basic improvements regularly every 12 months. So external design has been employed . . . to create an illusion of fresh values where none exist . . . When design is prostituted in this way, its own logic vanishes."

FROM TAILFINS TO BEETLEMANIA

In looking forward to a new design era based on something other than planned obsolescence, Teague observed that during the 1957 buyers' strike that sank the Edsel, Americans who had "cash and credit both went on buying other things." Of the things they bought most of, he writes, they mainly bought "small foreign cars as fast as they could get delivery, and especially they bought the little Volkswagen, about the smallest, most economical, sturdiest, and least pretentious of the lot . . . the Volkswagen scarcely changes its body style from one decade to another. In 1959, a total of 600,000 foreign cars [were] sold in the U.S.A. . . . [at the same time] exports of American cars . . . dropped sadly."

In the year that tailfins reached their peak of extravagance, the Volkswagen represented a sensible lack of pretense and a return to the same no-nonsense practicality that had put America on top before Sputnik. Of the 600,000 foreign cars that Teague tells us were sold in the United States in 1959, 150,000 were VW Beetles.

As early as 1956, *Road and Track* had marveled at the little car's ability to gain "an unmistakable wheel-hold in the garages and hearts of the American car-buying public." They also wondered, "How did it happen? Especially with practically no national advertising? . . . Probably the simplest [explanation] is that the Volkswagen fills a need which Detroit had forgotten existed—a need for a car that is cheap to buy and run, small and maneuverable yet solidly constructed . . . utterly dependable and trouble-free."

Volkswagen's lack of advertising did not last long. In Germany, VW was expanding its Wolfsburg plant facilities and would soon be producing more cars than ever before. Realizing that the six-month waiting list for Beetles in America indicated a tremendous potential market, Volkswagen appointed Carl Hahn as head of the newly created Volkswagen America in 1958. One of the first things Hahn did was to hire the tiny advertising firm of Doyle, Dane and Bernbach. In 1959 DDB was ranked eightieth among all American ad agencies. The creativity of their campaigns, however, had won them a lot of industry attention. In their earliest meetings with Hahn, partners Bill Bernbach, Mac Dane, and Ned Doyle decided that Volkswagen's unsung popularity was a natural reaction to Detroit's excess. They recognized that the Beetle was, as the cultural critic Thomas Frank would later observe, "the anti-car, the automotive signifier of the uprising against the cultural establishment."[43] In their war against the Detroit "dream cars," DDB decided honesty, irony, and humor would be the best way to expand Volkswagen's American market.

Other car ads of the late 1950s included beautifully air-brushed photos and mellifluously vague promises like: "Filled with grace and great new things," or "You ride in a wonderful dream car world of space and light and color."[44] By contrast, the Volkswagen campaign has been described as "the first time the advertiser ever talked to the consumer as though he was a grownup instead of a baby."[45] Because the product they represented was the polar opposite of Detroit cars, DDB positioned themselves directly against the idealizations of mainstream American car advertisements. Early DDB Volkswagen ads used un-doctored photographs and ran single-message sales captions like: "It won't drive you to the poor house," "Don't let the low price scare you off," or the completely subversive "Live below your means." Their success was palpable. In 1959 *Sales Management* observed that "the average American may be confused by Comets, Corvairs, Darts, Falcons, Hawks, Larks, Ramblers, Tempests, Valiants . . . But chances are he'll know one little bug by its changeless shape, and even know that its engine is rearward and air-cooled."[46]

As DDB's campaign for Volkswagen evolved, it fell right in with the emerging cultural idiom of cool and hip. Planned obsolescence soon became one of the primary targets of this idiom. DDB took death-dating to task in one of the best promotional campaigns in advertising history.[47] What they accomplished was really the first new development in marketing since Sloan's annual model change more than thirty years earlier. Essentially, Bill Bernbach, a master advertising strategist, solved the problem of overproduction and under-consumption by encouraging Americans to buy their product as an expression of their *rejection* of consumerism. In so doing, he established a continuous marketing trend. In 1961 a now-famous DDB ad ran the following caption under a photograph of a VW beetle lit by several spotlights on an otherwise darkened showroom floor. The single picture simultaneously introduced readers to "The '51, '52, '53, '54, '55, '56, '57, '58, '59, '60, and '61 Volkswagen."[48] DDB's point was obvious: Volkswagen did not make superficial model changes.

For the next decade, DDB elaborated their anti-obsolescence theme in print ads that attracted twice as many readers as other car ads, and also—according to Starch Company readership studies—scored significantly higher in readership surveys than any of the editorial content in the same magazine.[49] By the mid-1960s American magazine readers preferred a really good VW Beetle ad to almost any feature article. The ads reached a level of popularity that would not be matched until Budweiser developed its Super Bowl ad campaign in the 1990s. Advertising historians Charles Goodrum and Helen Dalyrymple described the attraction of the DDB print ads this way: "People stopped at the ads . . . read every word and were able to recall the illustration and the point months after publication." A famous VW ad made its anti-obsolescence point by running extreme close-ups of very minor features of a Beetle's body under the caption "How to tell the year of a VW." In the same ad, the box for the 1957 photo is blank and contains only the words "No visible change."[50]

My favorite is the 1961 caption that Bill Bernbach personally convinced Volkswagen to run at the bottom of a blank page. Instead of the usual photograph, DDB ran arresting uncluttered white space in a full-page magazine advertisement that probably cost the company $30,000. The caption made its anti-obsolescence point succinctly, although it also provoked surprised reaction at the parent company's offices in Wolfsburg. "We don't have anything to show you in our new models," read the DDB copy. Of course, the ad then went on to tell you why: Volkswagen did not believe in superficial styling changes to the body of the Beetle; anything new and noteworthy happened inside the car.

Gradually, by opposing the idealization and absurdity of Madison Avenue's consumer paradise (and especially the self-serving strategy of planned obsolescence), DDB put forward its own style of advertising as an antidote to the American establishment. Following the Volkswagen campaign, for a while, Americans bought fewer goods to keep up with the Joneses. Increasing numbers of Americans bought products like Volkswagens and, later, much more expensive Volvos "to demonstrate that they were wise to the game" and, ironically, "to express their revulsion with the artifice and conformity of consumerism."[51] In this way, Madison Avenue cleverly made the values of the counterculture accessible and acceptable to middle America, and then pressed them into the service of consumerism.

HATLESSNESS AND THE PEACOCK REVOLUTION

Cooptation of the emerging counterculture of the 1960s was most visible in the field of men's fashions. Myth has it that men's fashion liberalized suddenly in 1961 when Jack Kennedy appeared hatless at his inaugural address or when Pierre Cardin debuted the first full line of men's clothing by a major designer. In fact, Jack Kennedy *did* wear a hat to his inauguration, and although Pierre Cardin's line was remarkable, he was simply reacting to a variety of changes already visible in the market. The popularity of the T-shirt was one. Hatlessness was the other—a fashion that would eventually lead to the long hair we associate with the 1960s.

Until the 1950s, men's hats displayed their owners' perception of membership in a particular social class. But in 1954, in Elia Kazan's *On the Waterfront,* a hatless Marlon Brando was cast as the brother of Rod Steiger, a hat-wearing embodiment of the corrupt American male willing to sell out his underdog brother. Brando's hatless disenfranchisement stood in direct contrast to Steiger's hat-wearing corruption and his spineless conformity. The following year saw the release of *Rebel without a Cause,* in which James Dean played another rebellious and hatless role (originally intended for Brando). When Sloan Wilson's *The Man in the Gray Flannel Suit* explicitly connected the American male dress code to the stifling social conformity required of white-collar workers, hats were doomed to obsolescence, and gray flannel suits were not far behind.

In the mid 1950s, in other words, the rebellion in American male dress began as a slightly liberalizing change. At first it took a simple form—neglecting to wear a hat. Forgetting it, misplacing it, doing without it became a refreshing act of white-collar nonconformity. Eventually, these small acts would blossom into a men's wear revolution. Though at first many people failed to notice the hatless rebellion, the Hat Corporation of America was paying attention. In 1961 they ran a significant ad featuring a bearded, smoking, bespectacled, unsavory beatnik type, with the caption: "There are some men a hat won't help." The ad suggested that, for ordinary workingmen, hats would win the favor of their superiors and "make the rough, competitive road between you and the top a little easier to travel."[52]

But the Hat Corporation's ad was behind the curve. Already in 1957, an industry journal previously called *Apparel Arts* reinvented itself as *Gentlemen's Quarterly: The Fashion Magazine for Men*, three full years before Pierre Cardin made fashion history with his designer menswear. Before 1960 most men's clothing had a life cycle of five to seven years. A document written in 1959 by adman Henry Bach observed that "the industry has not found within itself the mechanism or the power to effect style obsolescence to the degree that it becomes self-generating."[53] But that was about to change.

Fashion advertisers were quick to apply DDB's Volkswagen strategies to the selling of men's apparel. The result was what *Es-*

quire columnist George Frazier would call "The Peacock Revolution." New colors and new fashions became available to American men more quickly than ever before. Buying and wearing these fashions were celebrated as "acts of rebellion" against gray-flannel constraint, conformity, and, by extension, consumerism itself. Madison Avenue found in the language of the anti-establishment revolution the very best means of encouraging lively repetitive consumption of menswear. A telling ad from 1968 illustrates the new advertising: "Men of the world arise! The revolution has begun and fashion is at the barricades. Charge in to Chapman's shops for men and lead the way to this new found freedom in men's clothes."[54]

Around this time James Rado and Gerome Ragni wrote and produced the musical *Hair*. "My Conviction," the spoken-song heard at the end of Act I, referred to the "flamboyant affections" of contemporary men's appearance, including their long hair. In retrospect, the lyrics were a manifesto for the Peacock Revolution. They described the "gaudy plumage" and "fine feathers" of the American male as a "birthright of his sex," and ended with the humorously scientific observation, "That is the way things are in most species."[55]

Two years after *Hair*'s opening night, menswear writer Leonard Sloane, in the *New York Times*, described the revolution in men's fashions as a marketing success story: "This trend toward obsolescence—as any customer who once bought a Mod Tie or a Nehru jacket must agree—is largely why the industry had record retail sales of $17.7 billion last year. And all indications point to more . . . high fashion merchandise with a short life span in the future."[56] Of course, once American manufacturers noticed that clothing was being sold by using the language of the counterculture, consumer products in many other fields followed suit. Anti–planned obsolescence advertising became a standard of the late 1960s. Even the corporate giants fell into step.

Perhaps because of this cooptation, the counterculture lasted well beyond the sixties and permitted Madison Avenue to engage in cycle after cycle of rebellion and transgression, marketing new goods, new fads, new symbolic gestures of defiance. The twist that DDB put on psychological obsolescence through their VW ads at the very beginning of the 1960s has been with us ever since.

THEODORE LEVITT AND MARSHALL MCLUHAN

Bernbach was not the only 1960s marketing genius who was obsessed with obsolescence. Fascinated by the economic theories of Joseph Schumpeter and Peter F. Drucker, Theodore Levitt, an oil industry executive born in Germany decided on a career change in the 1940s. After leaving his position at Standard Oil, Levitt completed a Ph.D. in economics at Ohio State University in 1951. By 1959 he had come under the influence of John Kenneth Galbraith and had joined the faculty of Harvard's Graduate School of Business Administration. An internationally respected economist, Galbraith brought the planned obsolescence controversy into academia by observing that

> a society which sets for itself the goal of increasing its supply of goods will tend, inevitably, to identify all innovation with additions to, changes in, or increases in its stock of goods. It will assume, accordingly, that most research will be induced and rewarded in the market place. Much will be . . . Under the proper circumstances . . . we may expect the economy to do a superior job of inventing, developing and redesigning consumer goods

> and improving their process of manufacture . . . Much of this achievement will impress us only so long as we do not inquire how the demand for the products . . . is contrived and sustained. If we do, we are bound to discover that much of the research effort—as in the automobile industry—is devoted to discovering changes that can be advertised. The research program will be built around the need to devise "selling points" and "advertising pegs" or to accelerate "planned obsolescence."[57]

Galbraith's *Affluent Society* was already leaving its mark on people who would soon become active in the young Kennedy administration.

In 1960 Levitt published the first of four highly influential essays in the *Harvard Business Review*. During his career as Edward W. Carter Professor of Business Administration at Harvard, these essays would each win a prestigious McKinley award and, collectively, would eventually earn Levitt the demanding job of editing the *Harvard Business Review*. The first of these essays, "Marketing Myopia," relied heavily on Levitt's experience as an executive for Standard Oil.[58] In a section entitled "The Shadow of Obsolescence," Levitt examined the myth of unimpeded "growth" in the most vital American industries. He demonstrated that even in a field as spectacularly successful as the petroleum industry, what most people perceive as unimpeded growth is really a "succession of different businesses that have gone through the usual historic cycles of growth, maturity and decay. Its overall survival is owed to a series of miraculous escapes from total obsolescence."

Levitt traced the development of the petroleum industry from its beginnings as a producer of patent medicines through a second period in which its major product became fuel for kerosene lamps. When lamps were rendered obsolete by Edison's light bulb, kerosene became a source of fuel for space heaters in American homes. But central-heating systems fueled by coal made the space heater obsolete as well. Just in the nick of time, the internal combustion engine came along. Levitt was careful to point out that obsolescence also created similar product shifts in a wide variety of major industries. In this first essay, his main point was that obsolescence dogs the heels of every manufacturer and that the product base of every business is subject to the "the usual historic cycles of growth, maturity and decay."

Levitt's observations concerning the obsolescence of Henry Ford's Model T were especially significant. He described Ford as the "most brilliant and most senseless marketer in American history."[59] Ford's contradictory behavior concerning the Tin Lizzie will be repeated in any industry, Levitt claimed, if "the industry has its eyes so firmly on its own specific product that its does not see how it is being made obsolete."

Five years later, in 1965, Levitt expanded these ideas in his second and most influential essay, entitled "Exploit the Product Life Cycle." In the phrase "product life cycle" (PLC), Levitt drew an analogy between Darwinian extinction and technological obsolescence. Every product, he said, goes through four specific life stages culminating in its decline or obsolescence (extinction). He showed that the inevitability of this cycle can be temporarily forestalled by strategies that promote the growth of the product's market. Levitt recommended three such strategies "aimed at" promoting demand among current users, new users, and "new market" users.[60]

Not surprisingly, this essay held enormous interest for American industrialists, manufacturers, managers, investors, and investment counselors. Eventually, Levitt's article led to widespread industrial awareness of the inevitable obsolescence of all products. In 1969, for example, Coca Cola Vice President Harry E. Teasley

Jr., an industrial engineer from Georgia Tech, applied Levitt's life cycle model in his groundbreaking cradle-to-grave study of the long-term energy expenditures involved in manufacturing glass versus plastic bottles. In 1972 this model was borrowed for a similar life cycle analysis of the milk bottle.[61] When the fuel crisis of the early 1970s struck, petroleum engineers used Teasley's study as a model for the first life cycle analysis of oil and alternative fuels. From Theodore Levitt onward, industrial design would increasingly take into consideration the environmental and economic impact of product obsolescence. His observations about the inevitability of product obsolescence marked the earliest beginnings of what would later be called "design for disassembly" or "green design."

The attraction of obsolescence as a topic of discussion led to one of the oddest phenomena of the 1960s: Herbert Marshall McLuhan. Patience, or impatience, but above all *interest* on the part of New York's advertising, artistic, business, educational, electronic, and media leaders turned this Canadian English professor, with his bad puns and gnomic utterances, into America's first, last, and only media guru. At the core of McLuhan's thought is the attempt to describe all of human history as a process of change in which successive media technologies rendered preceding modes of human consciousness obsolete. In McLuhan's world, obsolescence was more than just a favorite word. It was also the second of four fundamental laws that all media technologies throughout human history have obeyed.[62]

The Gutenberg Galaxy created a stir in 1962 by asserting that human society had already experienced three stages of consciousness and was now entering a fourth. The previous cultural turning point, McLuhan wrote, was the revolution of mind that followed the invention of moveable type. More than anything else, this invention was responsible for the individuation, specialization, mechanization, and visual orientation that developed during the reign of scribal culture after the fifteenth century. Print, McLuhan wrote, rendered oral culture (the preceding mode of consciousness) obsolete. He also believed that the electronic media of the twentieth century—telegraph, telephone, movies, radio, television, and digital computers—had brought mankind to the threshold of a new revolution in consciousness.

McLuhan's claim that "the medium is the message" reflects his conviction that whatever the surface content of a specific message, it is the technology of its medium that has the most lasting formative impact on the consciousness of human receivers: "If a technology is introduced either from within or from without a culture, and if it gives new stress or ascendancy to one or another of our senses, the ratio among all our senses is altered. We no longer feel the same, nor do our eyes and ears and senses remain the same."[63] The continuous reception of print created solitary and highly visual people with mechanistic tendencies. The emerging electronic media of the twentieth century was rendering obsolete the individualism and linearity characteristic of print culture.

In the period when he wrote and published his leading works, *Gutenberg Galaxy* and *Understanding Media* (1964), McLuhan famously claimed that the principal technology of scribal culture, the printed book itself, verged on obsolescence. For this reason, he wrote his books in a provocative, mosaic, nonlinear, and very difficult style that was "characteristic of electronic information movement" and that, he felt, was "the only relevant approach."[64] He borrowed the term "mosaic" from *Naked Lunch* (1959), mimicking William Burroughs's novelistic technique because it best reflected "the mosaic mesh of the TV image that compels so much active participation on the part of the viewer."[65] "Our planet,"

McLuhan wrote in a famous phrase, had "been reduced to village size by new media." A central characteristic of the global village created by television and computers was the "principle of simultaneous touch and interplay" whose chief characteristic is that "we are most at leisure when we are most intensely involved."

McLuhan's attempt to understand human activity and progress during a crisis of social change was mirrored in the work of European social theorists who called themselves post-structuralists. The measure of McLuhan's influence, however, was not in the robustness of his theories but in the unshakeable confidence with which he approached his enormous task. He was unique because he understood implicitly that Western attempts to construct meta-narrative accounts of human endeavor were constrained by linear, scribal thinking that television was then rendering obsolete.

Something (many people claim it was a brain tumor the size of a tennis ball, which was removed in 1967) momentarily freed McLuhan from conventional patterns of thought during the 1960s. In the span of a few years, his mad-hatter effusions drew followers to him from a wide array of disciplines and professions. As the recipient of a prestigious Albert Schweitzer fellowship at New York's Fordham University, he entered the American mainstream, where he alienated, befriended, or otherwise unsettled an impressive assortment of American cultural leaders, from Woody Allen, John Cage, Joseph Campbell, Buckminster Fuller, Abbie Hoffman, and William Jovanovich to Alan Kay, Stanley Kubrick, Timothy Leary, Victor Papanek, Ezra Pound, Carl Sagan, Tom Wolfe, and Andy Warhol.

By mid-century, McLuhan had become a recognized critic of advertising, like Vance Packard before him. McLuhan also had a ready explanation for Packard's popularity in criticizing the excesses of advertising. He felt that Packard as a social phenomenon had been made possible by the leveling effect of television, which allowed Packard to "hoot at the old salesmen . . . just as MAD [magazine] does."[66] The lively prose of a passage that slams General Motors and its ad agencies because they did not "know, or even suspect, anything about the effect of the TV image on the users of motorcars" is probably the best clue we still have to McLuhan's charismatic attraction. It is often claimed that he wrote messy, obscure books because he was a bad writer and a much better talker. As a devout fan of Lawrence Sterne and Salman Rushdie, I found McLuhan's writing very readable when I returned to *Understanding Media* during research for this book. Here is a sample, which provides a sixties perspective on that cultural moment in the late 1950s when Cadillac and its "scientific" admen went over the top:

> Only a few years back, Cadillac announced its "El Dorado Brougham" . . . We were invited to associate it with Hawaiian surf riders, with gulls soaring like sixteen-inch shells, and with the boudoir of Madame de Pompadour. Could MAD magazine do any better? In the TV age, any of these tales from the Vienna woods, dreamed up by motivational researchers, could be relied upon to be an ideal comic script for MAD. The script was always there . . . but not till TV was the audience conditioned to enjoy it.[67]

The obsolescence rate of the most advanced information is such that
within ten years about one half of the really important knowledge an
individual in these fields has acquired is outdated . . . A vice president
of one of our nation's three largest banks . . . put the useful life of the
knowledge of his technological staff at three years.

JOHN DIEBOLD, *MAN AND THE COMPUTER* (1969)

7   Chips

The 1960s saw the beginning of a shift toward an information society that would reshape the American cultural landscape. As abstract intellectual products reached the market—ranging from
disposable educational toys like Merlin, Quiz Wiz, Simon, and
Speak and Spell to video games, word processing programs, and
accounting spread sheets—obsolescence began to take on increasingly abstract meanings. Whereas in earlier decades the term applied strictly to physical objects (consumer products, the machinery that produced them, or the laborers whom the machines
rendered obsolete), in the 1960s it became possible to describe
people's knowledge, training, and skill sets as victims of obsolescence. By 1970 the technology writer James Martin would observe
that "the half-life of the technical training of computer personnel
is about three years. As more and more jobs and functions in
life become computerized, so the half-life of many persons' training will drop—in some cases to the three years of the computer
man."[1]

This new phase of obsolescence began with the appearance of increasingly powerful miniaturized circuitry in every corner of daily life, from the pinball parlor to the children's playroom. Together with transistor radios and hand-held calculators, these toys and video games mark the beginning of e-waste as we know it today. Paradoxically, though the size of computer hardware shrank exponentially, software applications proved steadily more difficult to replace or modify than progressively obsolete computer hardware. People proved less willing or less able to let go of hard-won skills than to dispose of the most expensive obsolete machines.[2]

EMULATION, MINIATURIZATION, AND MICROCHIPS

In 1964 IBM announced its System/360 line of mainframe computers. During the next five years, these machines proved so successful that IBM's sales more than doubled as the competition became increasingly obsolete. A cornerstone of the System/360's success was its ability to run the software applications of earlier, less powerful IBM computers without losses in processing speed. This capability was the result of a small revolution in computer architecture called microprogramming. Using the speed they gained through microprogramming, IBM engineers avoided the common difficulty of earlier attempts to imitate or simulate the applications of obsolete models of computers. This was known as the "Turing Tar Pit," named for Alan M. Turing, one of the founders of computer science. The phrase signifies a theoretical possibility that is extremely difficult in practical terms. In order to distinguish the System/360's dynamic processing feature from earlier attempts to imitate or simulate the applications of obsolete computer models, Larry Moss of IBM called this new ability "emulation."[3]

Talking up the advantages of emulation, IBM salesmen persuaded established customers to reinvest in System/360 hardware
by pointing out that their earlier investments in IBM software,
data storage, and personnel training would not become obsolete.
Model 65 in this line was especially popular because of its ability
to emulate applications for the 7070, the most popular large-capacity business computer. The 65 could run old applications up to
ten times faster than the machines for which the programs had
originally been designed. Realizing the success of this marketing
strategy, IBM devoted extra resources to their lower-end models
(Models 30 and 40), enabling them to emulate programs previously designed for the 1401.[4]

The first IBM System/360s were shipped in 1965, the same year
that saw the debut of the PDP-8 minicomputer, direct ancestor of
the personal computer. Among the germanium crystal transistors
of the PDP-8's architecture were the first integrated circuits ever
used in computers. Integrated circuits were essential to the personal computer revolution because they made compactness possible. Although the PDP-8 was still as large as an eight cubic foot
box freezer, Kenneth Olsen, one of the founders of Digital Equipment Corporation (DEC), the company that manufactured the
PDP-8, called his new product a minicomputer. He derived the
term from two British imports that were then enjoying considerable success in the United States—the miniskirt, and the Morris
Mini Minor, a small automobile whose ingenious design had
emerged in response to the Suez Canal crisis of 1956, which reduced oil supplies to Britain.[5]

Olsen knew that Morris had directed the famous automobile
designer Alec Issigonis to create a car that was lightweight, fuel efficient, and highly economical to operate. Similarly, compact integrated circuits would soon drive the computer revolution. The

PDP-8 would fill a market niche overlooked by IBM's behemoth mainframes. The Mini Minor also had another feature that appealed to Olsen as a product model: it consistently outperformed most of its overblown British competitors. In Gordon Bell, the principal computer architect for the PDP series since the 1962 debut of the PDP-4, Olsen had found a design genius with the same visionary zeal as Alec Issigonis. Bell's architecture took the opposite direction from that of mainframes. Bell believed that simpler machines with fewer instructions would consistently perform nearly as well as larger machines. The PDP-4 delivered over half the power of an IBM mainframe for about half the price ($65,000), and successive models consistently bettered that ratio.[6]

Because of the success of the compact PDP-8, DEC's revenues went from $15 million to $135 million between 1965 and 1970. Although the company was still only a fraction of the size of IBM, it was expanding at an unprecedented rate. IBM—content with 200 percent growth during this period—missed the significance of DEC's remarkable expansion, its corporate model, and its target market. Since the PDP-8 sold for only $18,000 and in no way affected IBM's market share, Big Blue executives were unthreatened by the new minicomputer. In 1970 DEC was selling as many minicomputers as IBM sold mainframes (about 70,000). By 1971, seventy more companies had formed to manufacture minicomputers. DEC would soon grow into one of IBM's major competitors, but last to realize this were the strangely complacent corporate ostriches at IBM.[7]

IBM's first pioneering breakthrough had been the use of transistors instead of vacuum tubes in their early computers. But the company had since developed a byzantine corporate culture that stifled innovation. For one thing, they seemed obsessed by mainframes—hulking, large-capacity computers modeled on the

UNIVAC whose complexity derived from the variety of processing speeds (and channels) they required for different functions. The huge mainframes IBM manufactured had little need for the miniaturization made possible by integrated circuits. The cumbersome design of mainframes emphasized strict, centralized control of computer data—a business model that reflected IBM's own bureaucratic ideology but was appropriate only for the largest corporations.

IBM's development of microprogramming in 1964 marked the company's last gasp of innovation. Although the sale of mainframes continued to grow for more than a decade, the writing was on the wall for any computer engineer who cared to read it. Small- to medium-size companies that could not afford to lease or buy from IBM's System/360 line looked to the PDP series to fill their needs. The limitations of the PDP-8—its inability to multitask, its short (12-bit) word length, and its relatively small (4K) memory—did not adversely affect these small users, who needed it for laboratory work, shipboard systems (especially in submarines, where space restrictions were a priority), office management, and inventory. Small local bank branches began to use the machines to handle their daily transactions before they sent updated records to a central mainframe at headquarters. This practice, called distributed computing, legitimized DEC's trend toward downsizing its computer architecture. The trend was further encouraged by the need for subminiaturized and lightweight circuits powerful enough to regulate the trajectories of Minuteman missiles and Apollo moon rockets.

In 1958 at Texas Instruments, Jack Kilby, a second-generation electrical engineer, was busy assessing a government-funded research project concerning subminiaturization using a device called the micro-module. By the end of World War II it had become clear that the future of aviation electronics depended on reducing the cost, size, and speed of electrical circuits while maintaining a high degree of reliability. A B-29 bomber had relied on nearly 1,000 vacuum tubes and tens of thousands of passive devices whose weight increased drag.[8] Transistors had presented an acceptable solution until 1957, when Sputnik I catapulted America into the space race and made smaller payloads a critical challenge. In 1958 the Pentagon eagerly began funding research into micro-modules because they promised a new level of subminiaturization by depositing printed electronic components on a ceramic wafer.

Although IBM would later use a similar device in the System/ 360s, Kilby thought that micro-modules were an inelegant and limited solution to the problem of subminiaturization. He also suspected these devices would prove expensive to manufacture. Instead, Kilby wondered if he could simply put a variety of small components on a single wafer of semiconductor material and connect them by embedding fine gold wires in the crystal. Semiconductors used germanium or silicon because these natural materials neither conducted nor resisted electrical current. Kilby correctly imagined that the cost of a germanium or silicon wafer would be offset by significantly lower manufacturing costs, since production, packaging, and wiring expenses would be limited to a single process. By the fall of 1958 Kilby had completed a working oscillator on a single wafer of germanium, and in 1959 he filed a patent for what Texas Instruments called "the solid circuit."[9]

From William Shockley, for whom he had worked at Shockley Semiconductor, Robert Noyce had learned that the first version of any innovation, such as the transistor or integrated circuit, is usually a crude device that can be quickly improved. Noyce later described Kilby's approach as "brute force"—"taking a piece of semiconductor . . . shaping it . . . and then . . . still doing a lot of wiring." His lab notebook entry from January 1959 records a scheme for creating a circuit similar to Kilby's germanium invention but doing it in silicon. Using the planar process invented by Jean Hoerni, a Fairchild employee, Noyce completed a design for a semiconductor circuit in 1959 that he called Micrologic. Noyce had already devoted a lot of thought to wiring and to what has been called the tyranny of numbers. Essentially, he recognized that wiring itself presented a problem to the success of subminiaturization because in addition to increasing the cost and weight of components, it also increased the distance that an electronic pulse had to travel: this in turn limited a given component's speed.[10]

What Noyce would call the "monolithic idea" evolved as a solution to the limitations presented by wiring. He used Kilby's idea of creating several components on a single wafer. To this he added Hoerni's planar process which had originally been intended to seal each silicon wafer with a layer of silicon oxide in order to prevent impurities like gas, dust, and stray electric charges from incapacitating a working transistor. Noyce's innovation also copied the basic idea of the micro-module. By printing a circuit's wiring directly onto the inner surface of Hoerni's silicon-oxide seal before applying it to the silicon wafer in which different components had already been created, Noyce completely eliminated the need for additional wiring. The idea was monolithic because it combined three cutting-edge technologies into a single sealed device. Noyce applied for a patent a few months after Kilby filed his own application, and it was granted in 1961.

Although Kilby's patent was refused and he actually lost the rights to his invention, by 1964 Texas Instruments and Fairchild reached an accommodation in which both men were credited for co-inventing the integrated circuit. By that time, the use of integrated circuits (ICs) was becoming pervasive. Still, IBM's administrators wanted nothing to do with them. As late as 1963 an internal memo answered the concerns of some young IBM engineers who worried that the solid-circuit technology (micro-modules) used in the designs and prototypes for the System/360 series would soon become obsolete. The memo stated that "monolithics" would not be a "competitive threat either now or in the next five years." Twelve months later, however, another memo noted that ICs had made rapid progress. This second memo also claimed that IBM was several months behind in this emerging technology and would require "six months to a year to catch up." Nonetheless, although ICs had become cheaper and more readily available, IBM continued to manufacture their System/360s using ceramic micro-module circuitry.[11]

Despite IBM's indifference, integrated circuits gained considerable acceptance elsewhere. In 1962 North American Aviation's Autonetics division won a lavish contract for the guidance system of a new intercontinental ballistic missile, the Minuteman II. Autonetics decided to take advantage of the subminiature ICs. The guidance system of the Minuteman I had contained over 15,000 discrete circuits. By the time of its first launch in 1964, the weight, size, and complexity of Minuteman II's guidance computer had been reduced to 4,000 discrete and roughly 2,000 integrated circuits. Between 1962 and 1965, the Pentagon signed electronics contracts totaling $24 million dollars. The half million chips sold in 1963 quadrupled every year until 1966. By that time, Autonetics was producing six new Minuteman II missiles weekly and calling for over 4,000 circuits a week from Texas Instruments, Westinghouse, and RCA. The Minuteman II program had become America's top consumer of integrated circuits.[12]

NASA was now paying attention, too. It had purchased integrated circuits since 1959 when Texas Instruments first made them available. In 1961 NASA charged its internal instrumentation lab with responsibility for the Apollo guidance system. Their administrator recognized that integrated circuitry was ideal for the Apollo guidance computer, of which seventy-five were built, each requiring about 4,000 ICs. These integrated circuits now came from several companies, including Texas Instruments, Philco-Ford, and Fairchild Semiconductor. Before Apollo, Robert Noyce at Fairchild had eschewed any involvement in military contracts like the Minuteman. The Apollo moon mission was a completely different matter, however.[13] NASA did not share the military culture that, Noyce felt, stifled innovation, promoted bad science, and championed limited solutions like the micromodule. NASA, he felt, was a unique scientific enterprise. Together, Noyce and Fairchild Semiconductor jumped eagerly into the Apollo moon mission.

As a result of joint patronage by the Minuteman and Apollo programs, semiconductor manufacturers dropped the price of their integrated circuits from $120 per chip to about $25 between 1961 and 1971. During this decade, the number of circuits that could be crammed onto a single chip increased dramatically.

This steady increase in the maximum number of circuits on a single chip had followed a predictable curve since 1959. The first person to notice this regularity was Gordon Moore, director of research at Fairchild Semiconductor and one of its eight co-founders (including Noyce). The Moore-Noyce friendship had begun when they met as young engineers working at Shockley Semiconductor Industries. Together with six of their fellows, Noyce and Moore left Shockley in 1957 over an ongoing management dispute. These eight men found, as had John Bardeen and Walter

Brattain, inventors of the first transistor, that Shockley was a less than ideal supervisor. Moore and Noyce moved again in 1968 when they became dissatisfied with Fairchild Camera and Instrument, their parent firm. This time they co-founded Intel, an independent semiconductor manufacturer that would become the industry leader for decades.

In October 1965 Moore published observations that would later become known as Moore's Law, and still later as Moore's First Law. Moore pointed out that the level of an integrated circuit's complexity had increased in relation to its minimum cost at "a rate of roughly a factor of two per year. Certainly, over the short term this rate can be expected to continue, if not increase. Over the longer term the rate of increase is a bit more uncertain, although there is no reason to believe it will not remain nearly constant for at least 10 years. That means by 1975, the number of components per integrated circuit for minimum cost will be 65,500. I believe that such a large circuit can be built on a single wafer."[14]

Although Moore's Law was intended to emphasize the increasing power and the diminishing costs of integrated circuits, it also provided an index to the steady rate of technological obsolescence created by ICs. In 1965 chips were doubling their capacity and lowering their price every year, so it did not take very long at all to render obsolete any given chip or the power, compactness, and cost of the device that used it. In other words, any electronic device that contained a microchip was death-dated by the time it left the assembly line. These devices were truly self-consuming artifacts, since their desirability diminished automatically. Every year, smaller and smaller electronic devices became available for less and less cost, and these devices became at least twice as capacious

and twice as fast as their immediate predecessor, effectively qua-
drupling the value of each generation of chip.[15]

The speed at which this technological obsolescence occurred
became obvious during Apollo's last flight in 1975, when American astronauts aboard this joint Apollo-Soyuz docking mission
carried with them programmable HP-65 pocket calculators that
were several times more powerful than the capsule's inboard computer designed less than a decade before.[16]

But Moore's Law is only part of the much bigger picture of the
history of computing, and the forces driving the acceleration of
obsolescence and e-waste are far older than integrated circuits.
In 1999 Raymond Kurzweil, winner of the prestigious National
Medal of Technology, observed that Moore's Law is actually "the
fifth paradigm to continue the now one-century-long exponential
growth of computing." Kurzweil pointed out that an

> Exponential Law of Computing has held true for a least a cen-
> tury, from the mechanical card-based electrical computing tech-
> nology used in the 1890 US census, to the relay-based computers
> that cracked the Nazi Enigma code, to the vacuum-tube-based
> computers of the 1950s, to the transistor-based machines of the
> 1960s, and to all of the generations of integrated circuits of the
> past four decades. Computers are about one hundred million
> times more powerful for the same unit cost than they were a half
> century ago. If the automobile industry had made as much prog-
> ress in the past fifty years, a car today would cost a hundredth of
> a cent and go faster than the speed of light.

(In June 2000, when Intel introduced a single-chip processor containing 150 million transistors, Moore would give the automobile
analogy an ecological twist, remarking that if automobiles had

improved at the same rate computers did, we would all be driving cars that got 150,000 miles to the gallon.)[17]

When Moore made his original observations in 1965, the cost of continuously replacing obsolete hardware was ameliorated by radical reductions in the price of newer, more powerful models. Technological obsolescence—the same market force that Walter Dorwin Teague had approvingly referred to as "natural obsolescence" in 1960—was driving the repetitive consumption of a variety of new products that now included digital watches, calculators, computers, and computer software. By 1965 the ground was prepared for America's e-waste crisis. The earliest e-waste product that contained a microchip was the disposable electronic calculator.

DEATH OF THE SLIDE RULE

Analog calculators, especially the slide rule, had dominated complex calculations since the early seventeenth century. In the years after ENIAC, large desktop alternatives became steadily available, but these electro-mechanical business calculators were unable to handle the size, complexity, or number of operations required by scientists, architects, and engineers. While PDPs and their successors made inroads into these communities, minicomputers were still prohibitively expensive for private or small-shop use. For this reason, complex calculations at the drawing board or in the lab were still being performed on a device that had been invented by William Oughtred in about 1625.[18]

The slide rule (or slipstick, as it was often called) had serious limitations. When John Atanasoff observed his students' frustrations in using slide rules to solve what he called "large systems of simultaneous algebraic equations for . . . partial differential," he began to contemplate a digital computer in 1935.[19] In order to distinguish between the capabilities and methods of the slide rule and those of the electronic computer he dreamed of building, Atanasoff began to refer to them respectively as analog and digital devices. By 1940 he and his graduate assistant, Charles Berry, had built a prototype digital computer, the Atanasoff-Berry Computer (ABC). John Mauchly would plagiarize Atanasoff's ideas in order to create his own large-capacity electronic calculator to handle weather data (see Chapter 5).

Despite its limitations, the slide rule had the advantage of being compact and readily available. By the early 1960s, large computers could satisfy most complex needs, but many people who needed such computing power still had very limited access. Elaine A. Gifford of the National Photographic Interpretation Center worked as a photo-grammetrist interpreting data in the top secret CORONA spy-satellite program. Despite their limitless funding, almost no one had adequate access to computer time. She remembers, "We didn't have hand-held calculators in 1965; during that period we had to look up trigonometry functions and use slide rules . . . The ground resources lagged behind the overhead satellite system." A few years later, an article in the *Electronic Engineer* entitled "An Electronic Digital Slide Rule" predicted that if it became possible to build a hand-held calculator, "the conventional slide-rule will become a museum piece."[20]

One solution to the increasing demand for calculating power and the simultaneous inaccessibility of computer time was to make desktop calculators more powerful and flexible. At Cal Tech a small programmable calculator went into the planning stages as early as 1966. In 1968, in Japan, Masatoshi Shima, an engineer at ETI, parent company of the calculator firm Busicom, had the idea of designing a programmable desktop calculator using integrated circuits containing 3,000 transistors, at a time when the most so-phisticated calculators used only 1,000. In 1969 ETI approached Intel with their design. Noyce and Moore assigned Ted Hoff, the company's twelfth and newest employee, to assist the Japanese in making a suitable set of components. Until that time, Hoff's main job at the company had been to find new uses for Intel memory chips.[21]

From the beginning, Hoff thought that ETI's complex design would prove too much for Intel's limited manpower. He also knew that a single memory chip was sufficiently large to store the pro-gram Busicom needed. Although Busicom had no need for a gen-eral-purpose computer, Hoff suggested that they could reduce their amount of logic (and the number of transistors) simply by using a memory chip to run calculator subroutines. At first, Shima and ETI's other engineers reacted negatively, but Hoff persisted. He next suggested that they put the central processing unit (CPU) of a simple computer onto a single chip and run it from stored programs on a few more Intel memory chips. Theoretically, this was possible, given the large-scale integration (LSI) of circuits on a single chip that had already been achieved throughout the semi-conductor industry. A microchip's capacity for integrated circuits had kept pace with Moore's Law, doubling each year since 1965.

The American's argument was compelling, and it soon con-vinced Shima, a skilled engineer who was intrigued by Hoff's ideas. When Noyce, the business genius of Intel, learned that Busi-com was experiencing financial difficulty, he offered them Intel's 4004 chip at a greatly reduced price, provided Intel retained world-wide rights to the Busicom chip. ETI agreed. This single deal would make Intel one of the most powerful computer companies in the world, while its Japanese partner would later be forced into bank-ruptcy shortly before Masatoshi Shima came to work for Intel.[22]

Hoff's microprocessor, the 4004 chip, relied on three other chips—two containing memory and another controlling input and output functions. Frederico Faggin, who did not share in the microprocessor's patent and soon left Intel to work at Atari, designed its complex circuitry. Busicom's printing desktop calculator, the 141-PF, was introduced to the Japanese market in April 1970. That same month, Canon and Texas Instruments introduced their Pocketronic programmable calculator to the Japanese market with circuits designed by TI's Gary Boone under the direction of Jack Kilby. Despite its name, the Pocketronic was actually a hand-held calculator weighing 2.5 pounds. Like its 1966 Cal Tech prototype (now on display at the Smithsonian), it was too big to fit in an ordinary pocket. In 1996 the U.S. Patent Office officially recognized Boone's "microcontroller." Although Hoff's 4004 Intel microprocessor had been invented earlier, Boone's was the first to combine input and output functions on a single chip.[23] Boone's TMX 1795 existed only in prototype. TI took its successor, the TMS01XX, to the production stage.

In 1971 just months after Canon and Texas Instruments introduced the Pocketronic in America, another Texas firm, Bowmar, unveiled the four-function 901B calculator, the first truly pocket-sized calculator. The Bowmar Brain, as it was known, was not very powerful, but it was entirely American-made and was also probably the first consumer device to use a light-emitting diode display, although LED alarm clocks and digital watches would not be far behind.[24] Bowmar simultaneously manufactured a second calculator, the C110, under contract to Commodore.

The next few years saw an avalanche of increasingly powerful, sophisticated, and cheap pocket calculators. Commodore introduced successive models, the Minuteman 1 and 2, between January and August of 1972. Intel released its 8-bit processor in April of the same year. By July, Hewlett Packard was using the new Intel chip in its first scientific calculator, the HP-35, which retailed for $395. With its capacity to perform logarithmic and trigonometric functions faster and more accurately than a slide rule or any other analog device, the HP-35 was a revolution for engineers and scientists.[25]

That year also saw the debut of the Aristo M27, a four-function calculator like the Bowmar Brain that used TI chips. It was a good basic machine, but the most significant fact about the M-27 was that its manufacturer was Aristo, Denner and Pape, a company that had mainly produced slide rules since 1872. Clearly, they were adapting to a changing marketplace. In 1973 TI brought out the first "slide rule calculator," the SR-50, which retailed for a mere $170. The next year initiated the calculator price wars. By January 1974 Aristo had introduced their own scientific calculator, the M-75, and by June Commodore had an entire line of calculators (the 700/800 series) for sale at around $25. Finally, in 1975 Aristo, Denner and Pape shut down slide rule production forever. Their main competitor, Keuffel and Esser, also stopped making slide rules and began selling calculators manufactured by Texas Instruments, using Gary Boone's TMS01XX chips.

Henry Petroski, a historian of engineering, recalled an ongoing debate among his faculty colleagues at the University of Texas in the early 1970s over whether students wealthy enough to possess a scientific calculator had an unfair advantage over their poorer classmates in tests and quizzes.[26] Following the price wars in 1974, this question became moot, and by 1976 a good calculator that had cost $395 in 1972 now cost less than $10. Calculator manufacturers were producing fifty million units a year, and competitive pricing had made them universally affordable. SR calculators, too, were becoming ridiculously cheap. In his final book, *The*

Green Imperative, Frank Lloyd Wright's most famous apprentice, Victor Papanek, shared this recollection from the 1970s: "One of my favorite photographs . . . showed more than 600 engineers' slide rules stuck into the ground around a neighbor's lawn, forming a tiny, sardonic, white picket fence. When I asked about it my neighbor's wife said, 'We bought these slide rules for one dollar a barrel . . . and used all six hundred.'"[27]

Of more interest than the diminishing cost of calculators and the demise of the slide rule is the obsolescence of the skill set that older-generation engineers possessed. Tom West and Carl Alsing recalled promising each other not to "turn away candidates" at Data General in 1978 "just because the youngsters made them feel old and obsolete." By the early 1980s it was hard to find a recent graduate or engineering student who still used a slide rule for calculations. Older engineers, on the other hand, were reluctant to part with them. A study by the Futures Group found that engineers in senior managerial positions universally kept their slide rules close by because they were more comfortable with analog devices.[28] The digital accuracy and speed that younger engineers took for granted meant less to those who had received their training before the 1970s revolution in calculation. Even Jack Kilby, the man who invented the integrated circuit and supervised development of TI's Pocketronic, preferred to use a slide rule. Sensitive to accusations of being dinosaurs, some old-school administrators kept their slipsticks under cover in desk drawers or cabinets and performed calculations on these obsolete devices in private.

Thus, by the 1980s, what younger engineers perceived as a democratization of calculation had in fact sheared the engineering world along generational lines. Age, not wealth, determined which engineers had the advantage. As the hacker culture would soon demonstrate, design and engineering were no longer the exclusive activities of a carefully trained elite. The term "obsolete" now applied both to the device that the older generation of administrators preferred and to the analog skills they used. By 1978 when James Martin wrote *The Wired Society,* nostalgic attachment to obsolete skills and devices had become a recognized phenomenon of the information age: "Old technology always has a momentum that keeps it going long after it is obsolete. It is difficult for the establishment to accept a change in culture or procedure."[29]

Gradually, during the late 1970s, calculator technology slipped off the cutting edge. Unit costs for calculators shrank to insignificance, and the brightest lights of the semiconductor firms moved on to newer challenges in and around Silicon Valley—at the Stanford Research Institute (SRI), Xerox-PARC, Apple Computers, and Atari. A quarter of a century later, when Palm Pilot inventor Jeff Hawkins left Palm to found Handspring, he described his decision as analogous to this shift in the calculator industry: "The organizer business is going to be like calculators. There is still a calculator business but who wants to be in it? They're cheap, and sort of the backwater of consumer electronics."[30] Palm's founder was contemptuous of calculators because they had become low-cost complimentary giveaways at the local credit union. Their cheap construction and short battery life sent them quickly into America's landfills, where they were unceremoniously buried—the first generation of microchip e-waste.

VISICALC AND APPLE II

Long before Palm's heyday, microprocessors continued to democratize calculation and render the skills and training of a generation obsolete. In 1971 Papanek wrote prophetically that "computers continue to take over (or . . . we relinquish to them) a greater share of those activities that we have heretofore thought of as exclusively intellectual—but which in fact are sheer monotony." Five years later, an MIT graduate in electrical engineering and computer science became weary of the endless number crunching required in his classes at Harvard Business School. Daniel Bricklin had already worked as a project leader and programmer for DEC's first word-processing program, WPS-8. In 1977, when the amount of raw math at Harvard got him down, he began to fantasize about a computer program that would liberate company presidents and accountants from the constantly shifting flow of numbers required to run a successful business. Before 1979, keeping a business ledger meant manually recalculating whole pages of figures each time a single variable changed. This very labor-intensive operation was a poor use of executive time. Bricklin later recalled, "I started to imagine . . . [a] word processor that would work with numbers."[31]

Until VisiCalc appeared commercially in October 1979, the market for the hardware and software of personal computers was confined to the narrow field of computer and electronic hobbyists. In April 1977 both the Apple II and the Commodore PET received an enthusiastic welcome from this small but growing market. August of the same year saw the debut of the universally affordable Tandy TRS-80. Still, prior to VisiCalc, there were few applications that appealed to a wide range of customers. Hobbyists themselves spent their application money mainly on computer games written at companies like Broderbund, Muse, On-Line Systems, and Sirius. Spreadsheets and word processing programs were exclusively confined to expensive mainframes or minicomputers, where computer time was very costly. Bricklin's spreadsheet program changed all that. It was a practical financial

tool with a tiny learning curve that displayed changes across a ledger page immediately, unlike mainframes. Using it, businessmen
became free to ask "what if" questions: What if we shave a penny
off our unit cost? What if we increase our production rate by ten
units a day? What if we offer our customers an additional 1 percent discount? What if we raise salaries by 3 percent? "It was almost like a computer game for executives," one industry analyst
has observed.[32]

VisiCalc was eminently affordable. At first the program retailed
for $100. In light of the enormous demand, however, VisiCorp
raised their price to $150 in 1979. In that year, an Apple II and
monitor, loaded with VisiCalc software, retailed for less than
$3,000. If their departmental budget could not be stretched to
cover the cost of a personal desktop loaded with VisiCalc, many
executives were willing to absorb the cost themselves. They walked
into computer stores by the thousands asking not for Apples per
se but for the machine that ran VisiCalc.[33]

Apple claimed that only 25,000 of the 130,000 Apple IIs sold
between 1977 and 1980 were purchased exclusively for the computer's ability to run VisiCalc software. Certainly the Apple II had
other features that made it an attractive machine, including an
easy-to-use color system and the Apple Disk II drive. But the term
"killer app" (application) was first coined to describe VisiCalc's
impact, and with good reason. In 1977 Apple sold 7,000 Apple
IIs. In 1979—the year of VisiCalc's debut—that figure jumped to
35,000. Most important perhaps was the word-of-mouth about
the Apple-VisiCalc combination that flowed from its financially
influential users outward. For the first time, a personal computer
had escaped from the hobbyists' market into the American mainstream. Apple II users trumpeted the low cost, ease of use, convenience, and general coolness of having an Apple II displayed

prominently on one's expensive wooden desktop. Small wonder that America's corporate elite were so enthusiastic.[34]

WORD PROCESSING AND WORDSTAR

But it was another killer app that would drive computers into virtually every small business office in America, and eventually into American homes. Word-processing programs had been available on mainframes and PDP minicomputers since the early 1960s, but typing and editing a document on a mainframe was a very inefficient use of valuable computer time. For this reason, one of MIT's first word processing programs for the PDP-1 was called Expensive Typewriter.[35]

Seymour Rubinstein founded his software development firm, MicroPro, in 1978 with the $8,500 he had in his savings account. His prototype word-processing system, WordMaster, appeared that same year. In 1979 Rubinstein unveiled WordStar, which was far superior to WordMaster because it enabled a small businessman with poor keyboarding skills to compose and edit a perfect document (a business letter, say) on-screen without retyping the corrections that had made White Out and the X-key feature of IBM Selectrics such huge sellers.

By the time WordStar appeared, computer hardware dedicated exclusively to word processing was available from Lanier, Vydec, Xerox, IBM, and Wang, but these machines required extensive training and a substantial financial commitment. Most large offices still found traditional typing pools to be more cost-effective. The ability to type 40 to 60 flawless words per minute was an essential and widespread clerical skill. Instead of relying on dedicated hardware, Rubinstein made WordStar available to executives who already owned Apple IIs. Then he went one step further by making a version available for down-scale users of Tandy's TRS-80, a computer that in the late 1970s sold for a mere $400. Copies of WordStar sold for a hefty $450, compared to the $150 retail price of the second version of VisiCalc. But despite this high price tag, between 1979 and 1984 nearly a million copies were sold—and because of it, MicroPro became a $100 million a year industry.[36] Together with its fellow killer apps VisiCalc and dBASE, WordStar eliminated much of the mind-numbing drudgery and delay that resulted from clerical and accounting work. Simultaneously, these apps empowered new users while rendering old skill sets—minute ledger work, the ability to type quickly and flawlessly—completely obsolete.

GRAPHICAL USER INTERFACE

The emergence of the graphical user interface (GUI) was part of the same trend toward the democratization of intellectual work enabled by personal computers (PCs). In this case, though, it was not calculating, accounting, typing, or record-keeping skills that were rendered obsolete but rather the high level of computer knowledge required to access the Disk Operating System (DOS) of any desktop machine. Until the Macintosh appeared, most users of personal computers had to know the inner workings of MS-DOS, the most ubiquitous operating system and the one used by all IBM-compatible machines.[37]

In *Computer: A History of the Information Machine,* Martin Campbell-Kelly and William Asprey explain the complexity of operating most personal computers prior to 1984:

> The user interacted with the operating system through a "command line interface," in which each instruction to the computer

had to be typed explicitly by the user, letter-perfect. For example, if one wanted to transfer a document kept in a computer file named SMITH from a directory called LETTERS to another directory called ARCHIVE, one had to type something like: COPY A:\LETTERS\SMITH.DOC B:\ARCHIVE\SMITH.DOC DEL A:\ LETTERS\SMITH DOC. If there was a single letter out of place, the user had to type the line again. The whole arcane notation was explained in a fat manual . . . For ordinary users—office workers, secretaries, and authors working at home—it was bizarre and perplexing. It was rather like having to understand a carburetor in order to be able to drive an automobile.[38]

Prior to the appearance of the Mac, only one company had simplified its user interface and greatly reduced the demands of the rigorous DOS learning curve for its operators. Between 1973 and 1975 Xerox Corporation designed and built a personal computer called the Alto at its Palo Alto Research Center (PARC). The Alto's interface was originally called WIMP, an acronym for Windows, Icons, Mouse, and Pull-down menus. At the center of its graphical user interface was the mouse, a device that had been invented in 1965 by Douglas Engelbart, head of the Human Factors Research Center, a work group that studied the Man and computer interface (or Mac) at the Stanford Research Institute. Engelbart based his invention on an obsolete engineering tool called the planimeter, an antiquated device once as common as a slide rule. When an engineer moved the planimeter over the surface of a curve, it calculated the underlying area. Engelbart's SRI team experimented with several other pointing devices, all of which were intended as alternatives to overcome the limitations of the standard *qwerty* typewriter. Engelbart later remembered how "the mouse consistently beat out the other devices for fast, accurate screen selection . . . For some months we left the other devices attached to the workstation so that a user could use the device of his choice . . . When it became clear that everyone chose to use the mouse we abandoned the other devices."[39]

In 1968 Engelbart and about a dozen helpers demonstrated the mouse in the now legendary Augmented Knowledge Workshop at the Joint Computer Conference in San Francisco. Many of his co-workers, including Bill English, who had completed the engineering details for the mouse prototype, later joined Xerox-PARC, where Engelbart's mouse was incorporated into a system of visual displays that included graphic symbols called icons.[40]

Alan Kay, another SRI veteran, joined PARC at about this time. While completing his Ph.D. between 1967 and 1969 at the University of Utah, Kay, along with Edward Cheadle, had built a computer called FLEX whose interface included multiple tiled windows and square desktop items representing data and programs. At PARC he focused first on the problem of managing the windows in which the mouse operated. At the time these were still the cumbersome tools envisioned by Engelbart's SRI group. They competed for space on the computer's monitor and did not overlap. It was very difficult for a user to keep track of which window he was currently working in. Kay borrowed his ingenious solution from the FLEX machine. He told his colleagues to "regard the screen as a desk, and each project or piece of project as paper on that desk . . . As if working with real paper, the one you were working on at a given moment was at the top of the pile."[41]

By 1975, using Kay's desktop metaphor, PARC had built about a thousand Altos, and Xerox's marketing group installed a small network of one hundred of these machines in the White House, both houses of Congress, a few universities, and some major corporations (including Xerox itself). From this position of prominence, Altos received a lot of favorable media attention—so much, in fact, that Xerox felt they had adequately prepared the American market for introduction of the Xerox Star workstation in 1981. Unfortunately, Xerox retailed the Star workstation for about five times the cost of a personal computer, the rough equivalent of an average person's annual salary. The Xerox Star's failure in the marketplace was as definitive as that of the Edsel in the fall of 1957.[42]

In 1979 Steve Jobs visited Xerox-PARC for a demonstration of their office-of-the-future network concept using the prototype Altos. Jobs was visibly impressed by these machines and their connectivity, which was facilitated by Alan Kay's program, SmallTalk. Jobs was astounded that Xerox was not yet selling Altos. "Why," he wondered, "isn't Xerox marketing this? . . . You could blow everyone away!"

The next year, Jobs lost control of the Lisa project at Apple, and after a bitter corporate battle he was reassigned to administer another project—a less powerful computer whose design originated with Apple architect Jef Raskin. Raskin had worked at SRI in the early 1970s when Engelbart's group was still focusing on problems with the "man and computer" interface. At SRI, Raskin also had extensive contact with the PARC personnel. Jobs now insisted that Raskin's new desktop computer should have features that were not in the original design, including Engelbart's mouse.[43]

Before he left Apple in 1982, Raskin named the new computer after a favorite variety of apple that grew abundantly in the hills around Cupertino. Raskin's name was really a pun, of course. To most buyers, a Mac computer was a variety of Apple computer, just as a Macintosh was a kind of apple. But in Raskin's computer-literate circle, the word Mac recalled the Man and computer interface that Engelbart's Human Factors group had been studying in the early 1970s at SRI. Cleverly, the name "Macintosh" evoked

212 | **Made to Break**

both the company that made the computer and the interface design problem that it solved.

For all their other differences, Jobs and Raskin agreed totally about their product's name. They jointly envisioned a computer targeted toward a large market of personal users with limited knowledge, and a budget to match. When Lisa debuted in 1983, it would cost $17,000. Although few at Apple were listening, Jobs knew in advance that Lisa's price tag would be a big problem. He did not want to make the same mistake with the Macintosh.[44]

Much of Lisa's high unit cost derived from its sophisticated combination of hardware and software. The machine's excellent performance relied on more than a megabyte of memory to run an elegant new operating system. Macintosh's designers pilfered Lisa's OS and rewrote it in greatly reduced machine code so that it would fit onto a single chip. The Macintosh project was cocooned in a separate building over which Jobs himself hoisted a pirate flag. John Sculley remembered that "Steve's 'pirates' were a hand-picked pack of the most brilliant mavericks inside and outside Apple. Their mission . . . was to blow people's minds and overturn standards . . . The pirates ransacked the company for ideas, parts, and design plans."[45]

Lisa turned out to be exactly the overpriced marketing disaster Jobs had predicted. Macintosh quickly became the company's only hope of survival, and Jobs regained enough influence within Apple to install John Sculley as CEO by late summer. As the elegant Macintosh approached its release date, Jobs got each of the forty-seven members of his team to sign their names inside the molding of the original design for the Macintosh case.[46]

Macintosh debuted in 1984, targeted to a family market, but it promptly fizzled. Its initial failure followed a remarkably cinematic one-time-only television ad that aired during the 1984

Super Bowl. Later, after the computer was repositioned, the Macintosh achieved a 10 percent market share primarily as a result of its use in desktop publishing and education. The easy learning curve of the Mac's intuitive GUI desktop made it ideal for use in the classroom among first-time student users who knew nothing about operating systems or command lines. Buying only one Mac per classroom made the computer a very affordable tool, and in this way Jobs' and Raskins' invention accessed and influenced an entire generation.

GUI had one more wrinkle. Microsoft, in the person of William Gates III, had designed operating systems for Apple since its earliest days. In 1981 Jobs' Macintosh group once again hired the software development giant to adapt some minor parts of the Mac's OS. Microsoft benefited in two ways. First, they could develop software applications for the Mac platform more successfully than they could for the IBM-PC platform because the latter involved competing from a position of inexperience with the likes of Lotus and MicroPro. Familiarity with the Mac OS enabled Microsoft to develop sophisticated spreadsheet and word processing applications (Word and Excel) that were first introduced for Macs but later adapted to the much more competitive IBM-compatible market.[47]

The second way Microsoft benefited was by gaining firsthand familiarity with the technology involved in developing a sophisticated graphical user interface. The Microsoft development project, originally called Interface Manager (later renamed Windows) actually began shortly after Bill Gates returned from viewing a prototype Macintosh at the invitation of Jobs in September 1981. Gates was clearly taken by the new Apple. Whenever he spoke of it later, his admiration was unabashed: "I mean, the Mac was a very exciting machine . . . You sat somebody down and let them use

MacWrite, MacPaint and they could see something was different there. We can say that was just graphical interface, but the Mac was the graphical interface."[48]

The inspired choice of the name Windows did not come from Gates himself. It was a strategic move by Microsoft's marketing division after Gates initiated the new interface project. Windows was deliberately chosen to achieve a brand identification that would become as successful as the word Xerox had become in photocopying. In order to compete with the cleverness of the Macintosh name, Microsoft's marketers decided to try "to have our name . . . define the generic."[49]

First released in 1985, version 1 of Windows was too greedy and too slow even for the new 80286 Intel processors. With over 110,000 instructions, Windows did not become practical until Intel produced its 386 and 486 chips in the late 1980s. In 1987 Microsoft released Windows 2.0, an interface that blatantly copied the look and feel of the Macintosh desktop. Apple sued Microsoft in early 1988, and the suit took years to resolve. By 1989, Windows-based applications had been introduced to the market and Microsoft had sold over 2 million copies of Windows 2.0. The Word and Excel applications that Microsoft had originally developed for the Mac platform were now making inroads into the IBM-compatible market. Microsoft had become the largest software manufacturer in the world.

By this time, the rapid pace of technological obsolescence was an accepted fact in software design. Application packages were updated every eighteen months or so, in a spiral of repetitive consumption. Users traded up for increasingly sophisticated software packages that took advantage of processing advances in speed and memory. By 1990 Bill Gates III had become the richest man in the world, and in 1992 Apple lost its lawsuit over Microsoft's adaptation of the Macintosh desktop. The complicated ruling established that Windows had been legally derived from the graphical user interface of Apple's Macintosh. Microsoft's Windows 3.0, another Mac-clone interface unveiled in 1990, contained 400,000 lines of code.[50]

Strangely, as more powerful processors, combined with Microsoft upgrades, encouraged consumers to trade in their old PCs for faster machines, the Macintosh demonstrated tremendous staying power. Apparently, elegant design itself was competing with obsolescence as a market force. Steven Johnson, a historian of computer interfaces, explained the Mac's popularity in these words:

> More than anything else, what made the original Mac desktop so revolutionary was its character. It had personality, playfulness . . . The Macintosh was far easier to use than any other computer in the market . . . It also had a sense of style. The awkward phrase "look-and-feel" popularized by Mac advocates reflects just how novel this idea was. There wasn't a word to describe a computer's visual sensibility because up to that point computers hadn't had visual sensibilities . . . Staring at that undersized white screen, with its bulging trash can and its twirling windows, you could see for the first time that the interface itself had become a medium. No longer a lifeless, arcane intersection point between user and microprocessor; it was now an autonomous entity; a work of culture as much as technology.[51]

Johnson was careful to assign credit for the Mac's personality to the original inventor of the desktop metaphor, Alan Kay. Many years before the Mac's launch, Kay—who had read McLuhan very carefully—came to the realization that the invention of overlapping windows had a sociological significance comparable to the invention of print. Computers, he realized, were a revolutionary new medium that had fundamentally changed "the thought patterns of those who learned to read." Here is Alan Kay's eureka moment in his own words: "*The computer is a medium!* I had always thought of [it] as a tool, perhaps a vehicle—a much weaker conception . . . [But] if the personal computer [was] truly a new medium then the very use of it would actually change the thought patterns of an entire generation."[52]

FROM PINBALL TO GAME BOY

The thought patterns of the generation Alan Kay altered with his desktop metaphor were simultaneously affected by another visual computerized medium: video games. Perhaps it was for this reason that in the early 1980s, Kay (briefly) occupied the position of chief scientist at Atari before moving on to become a research fellow at Apple Computers. As a student of McLuhan, he would certainly have been aware of the media guru's observations about games from *Understanding Media:* "As extensions of the popular response to workaday stress, games become faithful models of a culture. They incorporate both the action and the reaction of whole populations in a single, dynamic image . . . The games of a people reveal a great deal about them."[53]

Computer games had their beginning in 1961 when DEC donated a free computer—about the size of three refrigerators—to MIT. The complimentary PDP-1 met with puzzled expressions from many MIT faculty members, who were familiar with mainframes. It then fell into the hands of a group of graduate students, who taught themselves programming by designing software applications for the machine. Among them was Steve Russell, whose playful bent led him to create a game for his fellow programmers.

By 1962 he had completed a simple spacecraft shooter called Spacewar. This was the first computer game ever invented.[54] Its use expanded virally until, by the mid-1960s, there was a free copy of Spacewar on practically every research computer in America. Soon afterward, Magnavox put Odyssey, the first home video game system, on the market. Odyssey had over three hundred discrete parts, including dice, cards, play money, and plastic overlays to attach to a TV screen. These provided the backdrop for each of Odyssey's twelve games, which included football, tennis, and baseball.

Around this time, Nolan Bushnell, an engineer working at Ampex during the day, was spending his nights designing a Spacewar-style computer game. He sold it to an arcade game manufacturer, Nutting Associates, and promptly went to work for them. But the game, called Computer Space, was a flop because it was too complex, requiring a player to read a short manual before using the machine. Few arcade game players were willing to do that. Recognizing this limitation, Bushnell decided to design a game that would require no instructions. He left Nutting and formed a small company called Syzygy with two friends. Unfortunately, the word Syzygy was unavailable for trademark, so Bushnell resorted to a Japanese word, Atari, which means roughly check or en garde in the game of Go, which Bushnell loved to play.

Bushnell's business plan was to manufacture a coin-operated version of his new game and then sell it to a company like Bally Midway, the largest manufacturer of pinball machines in America. Together with his partners, Bushnell hammered together a working version of Pong in 1972. They set up a trial run in a bar called Andy Capp's in Sunnyvale, California. Bushnell then loaded another smaller prototype into a suitcase and left for Chicago, where he showed it to two of Bally's purchasing executives. They were unimpressed. Scott Cohen, who wrote the corporate history of Atari, explains that, "like other coin-operated-game manufacturers he had shown his electronic game to, [these Bally executives] were still in the electromechanical era, of which pinball is a prime example . . . Pinball companies make pinball bumpers, flippers, solenoids, relays, and mechanical scorers . . . Nolan was coming to them with an idea for a game with just two moving parts . . . There was nothing in Nolan's game that interested these executives—video games were simply not in the field of their experience."[55]

At almost the same moment that Bushnell was getting the brush-off from Bally in Chicago, the Pong prototype shut down suddenly after two days of frenzied use at Andy Capp's bar. When its designer, Alan Acorn, went to service the machine, he discovered it had simply jammed. The three-quart milk container he had installed inside the casing to catch quarters was completely full. Quarters had also backed up through the machine right to the coin deposit slot. People who had played the game at Andy Capp's the first night it was introduced had lined up at 10:00 a.m. the following day to play again.[56]

Heartened by the news, Bushnell started to make the rounds to distributors in order to secure advance sales and leases that would permit him to float a line of credit to manufacture Pong machines himself. At C. A. Robinson in Los Angeles, Bushnell talked to Ira Bettelman, who described how Atari's first game was viewed by the pinball industry:

> We were representing ten or twelve coin-operated game manufacturers and didn't think too much of Pong as a machine . . . Here comes something we know nothing about, weren't prepared

for and didn't know what to do with. It was a culture-shocking event . . . Regardless of our prejudices and fears of the unknown, we put the Pong machine out in the field with one of our customers and quickly found out that the returns were astronomical . . . The cashbox in the original Pong was a bread pan—a pan used to bake bread in—which held up to 1,200 quarters or $300. When the game was in a good location, this pan took about a week to fill.[57]

Pong's success sparked a wave of spin-offs, knock-offs, and rip-offs. Pinball manufacturers scrambled to get into arcade video games with any Pong derivative they could contrive. These included SuperPong, Pong Doubles, Quadra-Pong, Space Race, and Gotcha. *Fortune* magazine estimated that of the 100,000 Pong-style games manufactured in 1974, only 10 percent came from Atari, although other sources put this figure as high as 25 percent. In any case, pinball manufacturers, listening attentively to the voice of their inner greed, now entered the video era enthusiastically. Unfortunately, the arcade market for Pong burned out quickly as a result of these imitators, and Bushnell had to find another market for Pong. In 1974 he gave Acorn the task of designing a Pong game for the home market, then exclusively dominated by Magnavox's Odyssey, which had sold 100,000 units per year since its introduction in 1972. Each machine cost $100.

With the debut of Home Pong in 1975, Atari became the only game manufacturer to straddle both the arcade and home markets for video games. By the time of the annual toy industry show where it was first introduced, Bushnell had completely sold out his stock of games. Out of the blue, Sears Roebuck's most experienced buyer arrived on the doorstep of the modest Atari offices in Silicon Valley. Tom Quinn knew that Magnavox's Odyssey now represented a $22 million per year industry. He offered to buy every Home Pong game Atari could produce in 1975. When Bushnell told him that the company's production ceiling was a mere 75,000 units, Tom Quinn doubled the figure and arranged for financial backers so that Atari could enlarge its production line.

With two home video game machines available in an expanding market, coin-operated manufacturers might have realized that their environment was changing and they needed to adapt. But because the home video game market was essentially a subfield of consumer electronics, it went unnoticed by pinball manufacturers. After all, there was not yet any point of direct competition between home video games and electromechanical arcade games. Following Atari's example, major pinball companies like Bally, Williams, and Gottlieb had focused on producing video games for the market they knew best. No one thought to tap the home video game market. No one anticipated that, within the decade, home games would suck the client base out from under coin-ops or that a decline was coming in which Sega and others would dump their arcades and leave the American coin-op market forever.

By 1975 Bally Midway had moved quickly and adapted appropriately to what it understood as the challenge presented to the pinball arcade game market. They produced two very successful video games, Gun Fight and Sea Wolf. Sea Wolf especially was an overwhelming success, selling over 10,000 units. Then in 1976 General Instruments developed and released a new chip, the AY38500, that could accommodate four separate paddle games and two shooters. This six-in-one chip sold for less than $6 and promised to reduce production costs and retail prices for home video games in the very near future. Coleco became the first American company to use the GI chip in a home video game system. By the end of 1976 ColecoVision's sales reached $100 million dollars.

That same year electronics firms raced against newly formed start-ups to get into the home video game industry. In August 1976 Fairchild Camera and Instrument began to sell Channel F, the first full-color home video system and the only one to use replaceable game cartridges. Channel F used an advanced Fairchild-manufactured F-8 microprocessor and supplemented it with four Fairchild memory chips. Scott Cohen described the impact of Channel F on Atari sales in this way: "Channel F was competitively priced at $170. It made Atari's black and white dedicated game which played only Pong look as obsolete as a Brownie Box camera compared to a Polaroid."[58]

Moore's Law was beginning to have an impact in the home video game market. By 1976 the four-year-old Odyssey system and Atari's one-year-old Home Pong game were obsolete. So too was RCA's black and white Studio 2 system, even though it would not be shipped for another year, and even though it had the same replaceable cartridge feature as Channel F. What followed was completely unforeseen.

In 1977, confronted by too much choice and confused by the rapid rate of product obsolescence among home video games, consumers staged a buyers' strike. They stopped buying video games. Allied Leisure went bankrupt before it could deliver its machines. Magnavox quickly left the industry, absorbing the production costs of all unsold Odyssey games still in its inventory and canceling plans for a new four-player game. National Semiconductor immediately halted production of its improved Adversary gaming system. Fairchild withdrew Channel F machines from the inert market. The only home video game companies to survive 1977 were Coleco, which lost $30 million, and Atari, whose capital was mainly invested in its obsolete Home Pong inventory.[59] In Japan, however, where the market for home video systems was healthier, 1977 saw the beginnings of Nintendo.

In America, pinball was king again. The arcade game industry had unknowingly survived a year that threatened to strip it of its customers. The first computerized pinball game, Ralph Baer's Pinball, designed for release on the Odyssey system in 1978, never reached its intended market. In 1978 Bally Midway dominated the coin-operated arcade market with a combination of traditional pinball and its new line of video arcade games. Not until Bill Budge's Penny Arcade debuted in 1979 did pinball manufacturers fully realize that pinball's electromechanical days were numbered and that home video games posed a direct threat to their customer base. Atari itself cancelled development of their pinball division under the new management of Ray Kassar in 1979. The years that followed would see several computerized pinball games and a simultaneous renewal of 1940s-style moral campaigns against coin-operated pinball machines and arcade culture in general.[60]

The same years saw a pleasant and readily accessible alternative to arcade games in new home gaming systems from Atari, Sega, and Nintendo. In 1985 the coin-operated arcade market collapsed as completely as the home video market had in 1977. By this time, video games were ubiquitous. Coca-Cola had even unveiled a series of vending machines with built-in screens allowing customers to play video games after they had purchased a soft drink. By 1983 Nintendo's first portable video game product, Game & Watch, a small pocket-calculator-size gaming device that also told the time, had sold millions of units (many of them counterfeits) throughout Asia. In light of its success in the Pacific Rim, Nintendo decided to export Game & Watch to the United States, but after substantial losses Nintendo quit distributing them in 1985. In 1988, with the introduction of an improved Nintendo Entertainment System (NES), the manufacture of expensive electromechanical pinball machines declined substantially, and in 1989 Nintendo delivered the critical blow.[61]

Game & Watch grew out of Nintendo engineer Gunpei Yokoi's observations about the failure of Milton Bradley's Microvision, launched in 1979. Microvision, a programmable handheld gaming device, had enjoyed little support from its parent company. Although initially MB sold over eight million units of Microvision, there were few game titles available, and only two or three of them were released before the machine's demise in 1981. Learning from this mistake, Game & Watch combined the features of low unit cost with great variety. Between 1980 and 1986, Nintendo itself released fifty-nine Game & Watch titles. Game & Watch enjoyed overwhelming success in Asia, with sales exceeding 40 million units. Nintendo was anxious to repeat this success.

Game & Watch was simple and unsatisfying to play. What Nintendo really needed was a pocket-size gaming system that played sophisticated and changeable games. Yokoi set to work on a new programmable device that maintained the distinctive control-cross or D-pad of Game & Watch—which would become the industry standard for all video game consoles. This time, using everything they learned from the Game & Watch experience and without waiting for a separate American launch, Nintendo introduced Game Boy to a worldwide market. Suddenly, sophisticated video games were portable, and this single move threatened to make both arcades and home systems obsolete. In a series of clever negotiations and subsequent lawsuits, Nintendo kept its American and European competitors from using the world's most popular video game, Alexey Pajitnov's Tetris. Nintendo then released a flashy, updated Game Boy version. David Sheff, author of a corporate history of Nintendo, wrote, "There is no way to measure accurately how much 'Tetris' contributed to the success of Game Boy . . . Once a customer bought one, Nintendo could sell more games, an average of three a year at $35 a pop. Not counting Game Boy, 'Tetris' brought Nintendo at least $80 million. Counting Game Boy, the figure is in the billions of dollars."[62]

Gradually, the largest pinball manufacturers—Bally, Williams, and Gottlieb—merged or were sold to WMS, a larger electronic concern. In 1992, 100,000 pinball machines were manufactured in the United States, but by the year 2000 that number had shrunk to 10,000.[63] Still, pinball was tenacious. Unlike Game Boy, it allowed physical movement, and it reinforced the player's pleasure with bright lights, lots of noise, and good thumping mechanical vibrations. WMS and another company—Stern, of Melrose, Illinois—explored a new market among aging retro users who could afford the high unit cost ($3,500–$7,500) in order to enjoy the game at home or in a club. Gradually, however, this market shrank also. From 1997 to 2000, WMS sank $18 million dollars into their pinball division, and by 2000 they were losing $1 million a month. That year, citing declining interest and prolonged losses, WMS announced its reluctant decision to end pinball production forever and to concentrate on coin-operated gambling machines.[64]

The only company in the world that still produces pinball machines is Stern, which makes a few thousand machines a year for the retro market, most of which is in Europe. In 2002 an *Economist* article entitled "The Last Pinball Machine" described pinball's obsolescence using the language of evolution. Pinball, the *Economist* wrote, is "now a species close to extinction." The article gave the reason for pinball's demise as generational: "Youngsters, raised on PlayStations and 3-D virtual combat, no longer feel like leaning over a glass cabinet and batting around a metal ball." Pinball and video games provide very different kinds of experience, appropriate to styles of consciousness from different eras. Pinball is tactile, mechanical, and physically labor-intensive, like Vannevar Bush's differential analyzer. The pace of video games, as Gary Cross points out, is "set by the game's electronics, creating an experience of great emotional intensity."[65] It is this emotional intensity, and the dopamine that floods players' brains during the game, that produces the potential for addiction and abuse.

Christopher Geist, a professor of popular culture at Bowling Green University, observed that video games have become a social phenomenon of major proportions: "I don't expect anyone ever expected video games to have such a fundamental impact on our society in so many areas. [They] have become an integral part of the fabric of American life, changing the way we think, the way we learn, and the way we see the future."[66] Nolan Bushnell, the founder of Atari, agrees and offers an informed description of the new consciousness shared and enjoyed by our gaming children: "In the future, in part, we will be living in virtual reality . . . To survive and make it in that dimension, we are going to have to be mentally awake. We are going to have to live and be comfortable and maneuver in a computer environment. These kids are in training." John Perry Barlow, founder of the Electronic Frontier Foundation, puts it slightly differently. He refers to the emerging generation of gamers as the new "natives" of cyberspace.[67]

NSC Official Gus Weiss used [the Farewell] . . . material to design a massive deception program . . . unparalleled in the cold War. With the collaboration of the CIA, the FBI, and the Pentagon, products were made, modified, and made available to Line X collection channels. The products were designed to appear genuine upon receipt but to fail later. Line X operatives and Soviet manufacturers blamed each other for faulty collection efforts or for the inability to copy correctly the blue prints . . . It was a blow to Soviet military buildup.

NORMAN A. BAILEY, *THE STRATEGIC PLAN THAT WON THE COLD WAR* (1999)

8 Weaponizing Planned Obsolescence

The development of the integrated circuit provided American manufacturers with a substantial and threatening lead over the Soviet Union in a technology vital to defense. In addition to innumerable industrial applications, miniaturization and the high-speed computations made possible by ICs were essential to the design of successful ICBMs, manned rockets, airplanes, and satellites. Yet from 1970 until its collapse in 1991, the USSR lagged ten years behind the United States in the manufacture of workable chips. In retrospect, it should surprise no one that the Soviet defense and security machinery were desperately pressed into the service of the state and charged with using any means to procure these essential devices. What is remarkable is that the KGB failed so spectacularly in fulfilling this mission.

The year 1970 saw the debut of the Pocketronic calculator in Japan and the publication of *Design for the Real World* in America, in which Victor J. Papanek condemned the wasteful practices that sustained America's throwaway culture. For the first time, a connection was drawn between America's domestic consumer practices and global geopolitical reality: "When people are persuaded, advertised, propagandized, and victimized into throwing away their cars every three years, their clothes twice yearly, their high fidelity sets every few years, their houses every five years, then we may consider most other things fully obsolete. Throwing away . . . may soon lead us to feel that marriages (and other personal relationships) are throw-away items as well and . . . on a global scale countries, and indeed subcontinents are disposable like Kleenex."[1] But 1970 was notable for another event underappreciated at the time. In that year the brilliant cold warrior responsible for turning planned obsolescence into a weapon against the Soviet Union joined Richard Nixon's staff as an economist.

Although Gus Weiss fulfilled his economic duties meticulously at the White House, his other interests proved much more geopolitically significant. Weiss had been fascinated with technology since he was a boy, when, during World War II, he developed an interest in aeronautics. Instead of fading in adulthood, Gus's fascination with airplanes became obsessive. On a break from his MBA studies at Harvard, Weiss took a road trip to Cape Cod with Richard Eskind, a good friend. When they passed a small Massachusetts airfield, Gus suddenly shouted "Stop!" and jumped out of the car. He disappeared for an hour, and the next time Eskind saw his face was through the window of a single engine craft as the plane came in for a landing. By way of explanation Weiss simply said, "I wanted to see what Cape Cod looked like from the air."[2]

In later years, Weiss would be decorated by NASA for his assistance in tracking a disabled nuclear-powered Russian satellite. He would also receive recognition by France's Legion of Honor (the first of two such recognitions) for his assistance in facilitating a joint project between General Electric and Snecma—the Société Nationale d'Etude et de Construction de Moteurs d'Aviation. This partnership created the CFM56 jet engine, the single most popular aircraft engine ever built and the prototype engine for Airbus. Weiss's meetings with French officials would provide him with a wonderful opportunity to take the Concorde on a demonstration flight from Washington to Paris in 1973. Following his second supersonic flight, Weiss gushed about the Concorde, "It spoils you rotten."[3]

Gus's lifelong Nashville friends recalled that many normal childhood events had been painful for him, perhaps because his expansive intellect made him a "real oddity" in the Weiss household.[4] Set apart by extraordinary intelligence at an early age, Weiss would certainly have followed world events in the pages of the *Tennessean* when he entered adolescence in the mid-1940s. Loathing Nazi Germany, he wore a Navy surplus bomber jacket during his high school years, sometimes well into the month of June. He shared the apprehensions of other Jews during the era of the Red Spy Queen scandal and the trial of Julius Rosenberg—events that helped shape his dogged patriotism. Weiss would go on to become a vigilant and pragmatic Soviet watcher who never forgot that the USSR had been aggressively spying on America since World War II.

After completing an economics degree at Vanderbilt, Weiss studied for an MBA at Harvard and later taught economics at New York University. In the mid 1960s he joined the Hudson Institute and became a colleague of Herman Kahn, himself a cold war original who had "thought about the unthinkable" and put it all down in his chillingly popular analysis, *On Thermonuclear War* (1961). Kahn was one of two men who inspired Stanley Kubrick's character Dr. Strangelove in the film of the same name (the other was Edward Teller).

After Nixon became president in 1969, Weiss joined the White

House's Council on International Economic Policy. Having few friends with whom he was truly intimate and trusting, he compartmentalized many aspects of his life. The students he mentored in evening classes at Georgetown University were later surprised to learn that he had been working at the White House for the National Security Council. His closest and oldest Nashville friend knew nothing of his most significant romantic involvement. A wider group of acquaintances in Washington enjoyed his company but saw only carefully selected facets of the larger man. One in this circle described him affectionately as a "fat, pink cherub," while another, looking a little deeper into his offbeat heart, described him as a person who "hid his torments well."[5]

One of these torments was a condition Gus had developed sometime soon after high school: alopecia universalis, complete and permanent loss of body hair. Physically, this condition made him unusual. Psychologically, the disease left Weiss permanently shy about this body, and added an extra dimension of difficulty to the problem of finding a suitably brilliant woman to share his eccentric life in the shadows of Washington. But by middle age, Weiss had adapted reasonably well to the black world of intelligence and the gray world of analysis. He was known for his brilliantly associative mind and truly twisted sense of humor, as well as his ill-fitting toupee, affection for dogs, and vicarious enjoyment of the family life of others.

At some point during his early Washington years, Gus Weiss created an informal group of about twenty like-minded professionals from different corners of the intelligence community who met often to share drinks, dinner, and information. Fancifully, they called themselves the American Tradecraft Society.[6] This organization provided Weiss with the social dimension that had been missing from his life since the days of fraternity dinners at Vanderbilt. But American Tradecraft was also designed to fill a very serious function with respect to national security. As a trained economist and avid student of history, Weiss was aware that one of the hottest topics among contemporary European economists in the 1970s was the role that espionage had played in the expansion of industry in early modern Europe.[7] Weiss saw parallels between France during the Industrial Revolution and the Soviet Union during the Cold War—particularly the imbalance in the overall number of inventions that had existed between France and England on the one hand and that currently existed between America and the Soviet Union on the other. Weiss felt that in both economic contests, as a result of this imbalance, technology transfer between nations had become imperative, and industrial espionage was therefore inevitable.

Although Weiss had no concrete proof of espionage, he viewed the USSR's rough military parity with suspicion, since many Soviet innovations mirrored their American counterparts, and Soviet funding for research and development was disproportionately low. Others were beginning to notice these discrepancies as well. A specialist in the economic aspects of Soviet technology recalled that "the technology transfer issue was hot in the 1970s and into the 1980s. I got interested in it in the early 1970s when I worked temporarily (in Moscow and London) for the Foreign and Commonwealth Office. It was then beginning to be talked about as an important issue in East-West relations. Initially there were lots of extreme claims—that the Soviet system was so hopeless that it couldn't absorb anything from the West, or that in both civilian and military technology we were giving away the crown jewels."[8]

By the time Weiss took his first position at the White House in 1970, he was already reasoning that the cash-poor Soviets were continuing their postwar strategy of sparing themselves weighty R&D costs by simply stealing or covertly buying the American technology they needed. As an economist, he knew that the arms race drew its strength from technological parity, as successive weapons systems rendered older systems dangerously obsolete and vulnerable to first strikes. Since Sputnik I, American military strategists had bet on the vitality of the American economy against the costly national restructuring required in a Soviet Union devastated by World War II. The Pentagon believed that by developing newer, more sophisticated, and more costly weapon systems as often as it could, the West could force the Soviets to overextend themselves economically. The USSR would reach a point where it simply could not keep up and would have to choose between aggressive military imperialism and sustaining its costly civil infrastructure. At its core, this was a strategic extension of the policy Eisenhower had expressed in 1956 when he committed America to continuously upgrading the technology involved in national defense: "In these days of unceasing technological advance, we must plan our defense expenditures systematically and with care, fully recognizing that obsolescence compels the never-ending replacement of older weapons with new ones."[9]

This is how repetitive consumption and planned obsolescence became one of the mainstays of America's geopolitical strategy. Ironically, this strategy engaged the Soviet Union in the purist kind of capitalist venture—a competition to produce a better product, in this case a more efficient, better-defended society. But the irony of this contest soon faded as the arms race, and its deadly potential, became the main focus of the cold war. In hindsight, we can see the maddening excesses of the military buildup from the 1950s through the 1980s, along with the space race, as a kind of Strangelovian strategy to spend the USSR into oblivion. Viewed from this perspective, the Soviet Union was doomed to play a loser's game of catch-up as the technological gap widened over the course of the cold war.

INDUSTRIAL ESPIONAGE

Earlier than most people, Gus Weiss observed that, although the Soviets were struggling, they were also somehow keeping up. Increasingly, he felt that they were doing this mainly through theft and deception. In 1971 an entire cadre of Soviets spies had been expelled from England for espionage activities focusing on science and technology. When the Russian Oleg Adolofovich Lyalin became a British agent in February of that year, he confirmed much of the information MI6 had already garnered from other informers. But Lyalin also disclosed at least twenty new names of Soviet agents engaged in industrial espionage in the UK. Some of these had begun their careers as techno-bandits as early as World War II. Infuriated, the British government decided to clean house; 105 Soviet citizens were expelled from the country in September of 1971.[10]

Eventually, it became clear that the Soviet agents sent home included many of the most influential figures in the KGB's Directorate T, which was staffed by an elite group of specialists called Line X. Their exclusive function was to acquire Western technology for use in the "reproduction" of Soviet "fatherland analogues." Much of what was called "research and development" in the USSR was actually a massive program of reverse engineering, based on stolen or covertly purchased technology.[11]

Before this espionage information became available, however,

Gus Weiss began to collect stories about covert Soviet attempts to acquire technology in the United States. Alone in his Washington office, ten years before the name Line X was uttered outside a small circle of Soviet operatives, Weiss conjectured that there must be a concerted, ongoing, and centralized Soviet effort to steal Western technology. He took his ideas to the security establishment, but CIA officialdom was uninterested in home-cooked intelligence, even by such an intelligent outsider.[12]

The meeting was not entirely fruitless, however, since Weiss was soon contacted by Helen L. Boatner, a CIA officer who then managed the agency's operations center and would later receive the Distinguished Intelligence Medal. In 1973 Boatner became a founding member of American Tradecraft, risking her job to tell Weiss that a group of Soviet scientists had obtained visas and permission to visit the Uranus Liquid Crystal Watch Company in Minneola, New York. Then, three days before their visit, according to Boatner, the group had used a loophole in American visa regulations to "expand" their approved itinerary to cutting-edge American semiconductor firms, including IBM and Texas Instruments. Under CIA surveillance, these visiting Soviet scientists then scoured the clean rooms of the American semiconductor giants with cellutape attached to their shoes, picking up trace material samples as they went. It was clear to both Boatner and Weiss that someone in Moscow had figured out how to take advantage of U.S. regulations by first gaining approval for the scientists to tour an innocuous firm and then changing their itinerary at the last minute to visit much more sensitive targets.[13]

Boatner's trust in Weiss, at considerable risk to her own career, highlighted an interesting facet of Gus's personality. His ability to connect on a personal level and win the immediate respect of the brightest and most committed professionals was his greatest strength. Though obviously weird, he had a self-mocking charm that made him unthreatening and even quite likeable (he regularly referred to himself as Gus Mitty Weiss or as Dr. Strangeweiss). Mostly, however, his transparent brilliance and invaluable frankness outweighed his oddity and secured his position with American Tradecraft's mid-level intelligence and security personnel, a group deeply committed to their jobs and much more concerned with effective counter-intelligence than with the style or strictures of official policy. Weiss's formidable gift for thinking outside the box was an extremely rare commodity among the Harvard, Yale, and Princeton–trained professionals inhabiting the upper echelons of the intelligence community. Famous for his belief in competitive intelligence analyses, William J. Casey, the future CIA director, always listened very carefully to whatever Gus Weiss had to say.[14]

Harris A. Gilbert, Weiss's boyhood friend, fraternity brother, and personal attorney, was present at the Washington ceremony marking Weiss's induction into France's Legion of Honor. He recalled what the young adviser sitting next to him said: "Gus is the only person in the White House that we can trust. He has no personal agenda, has no political aspirations, he tells you exactly the truth, and when you want to have your ideas . . . [and] analyses tested, he is the one who will listen to you and give you good advice."[15]

Despite his highly unorthodox approach to building a career, Weiss's circle of influence in Washington began to expand. As American Tradecraft dinners grew in size, Soviet efforts to acquire U.S. technology became more and more apparent. Without any budget or official policy in place, Tradecraft members initiated unofficial blocking maneuvers that became increasingly sophisticated and collaborative. Leaving no paper trail, this small group of friends operated invisibly until well into the 1980s, when they were finally penetrated by an East German informer.[16]

By 1974 the Nixon White House had officially put Weiss in charge of problems associated with technology transfer. Largely as a result of his recommendations, National Security Decision Memorandum 247 was issued later that year, prohibiting the sale of powerful computers to the USSR. This was the government's first attempt to take control of the transfer of American technology, and so it marked a turning point. Previously, Secretary of State Henry Kissinger had softened trade restrictions with the Soviets in a sincere effort to achieve détente. Following the 1972 summit, the flow of supercomputers to Russia had increased. But giving the Russians—who were, and still are, incapable of making cutting-edge integrated circuits—unlimited access to such vital technology proved to be a strategic mistake.

Some of the implications of that mistake became clear in Afghanistan. Responding to Kissinger's détente initiative, Chase Manhattan Bank opened a Moscow office in 1973 and agreed to finance the Kama River truck plant, which used IBM System 7 computers to regulate production, especially the automated forging equipment necessary to cast engine blocks. Despite Soviet assurances that these trucks would never be used for military purposes, the USSR's invasion of Afghanistan six years later was facilitated by an armada of Soviet vehicles whose drive trains included Kama engines.[17]

In 1981 Weiss had been at work on technology transfer to the Soviets for about a decade when President Francois Mitterrand invited President Ronald Reagan to meet with him privately en route to the G7 Summit in Ottawa. The French would soon pass a wealth of Soviet documents into American hands, and considering Weiss's portfolio with the NSC and his Legion of Honor decoration, it was inevitable that this formidable body of intelligence—now known as the Farewell Dossier—would eventually come into his possession.

By all accounts, Reagan had no reason to expect any such gift—and every reason to be suspicious of France's new leader, who was a lifelong socialist and whose coalition cabinet included four communist ministers. But Mitterrand wanted a close and trusting partnership with America and had no great love for the Soviets. In particular, he viewed the Urengoi pipeline that was to provide Western Europe with Siberian natural gas for decades to come as a very mixed blessing, one that would guarantee France a dependable and fairly low-cost energy source but would also make the whole of Europe more dependent on a Soviet Union strengthened by hard European currency. Mitterrand had little interest in increasing the power of the USSR, and he had considerable interest in expanding France's own international influence.

When Mitterrand asked for and was granted a private meeting with his American counterpart, he told Reagan that late in the previous year a completely trustworthy French citizen had informed French domestic counter-intelligence (the Direction du Surveillance du Territoire, or DST) that a well-placed KGB insider was offering his services to France. This agent, to whom the French gave the misleading English code name Farewell, had already been contacted in Moscow, and he had now passed into DST hands more internal Soviet intelligence documents than had been available since the 1960s, when Colonel Oleg Vladimirovich Penkovsky of the GRU (Soviet military intelligence) gave 5,500

Minox exposures to the SIS (now known as M16), including vital revelations about Soviet ICBM disposition and chemical warfare strategies.[18]

Mitterrand told Reagan that the Farewell documents contained startling revelations. Chief and most welcome among these was confirmation that not only was the Soviet Union desperately behind the West in its technology but that the USSR did not have sufficient economic dynamism to sustain any real research and development program of its own. As a result, the Soviets were compelled to steal whatever they needed in order to patch large technological gaps in every area of their country's infrastructure. In particular, at that moment the Soviets were desperately concerned with acquiring oil and gas pipeline technology in order to develop the westernmost Siberian oil fields, a huge and desperate effort that in cold war terms had taken on Manhattan Project dimensions.[19]

Mitterrand emphasized that if the Russians could complete their plan for this double-tracked trans-Siberian natural gas pipeline, the cash flow would enable them to strengthen their economy, pay off their debt, develop their infrastructure, and at last achieve military parity and perhaps victory in the cold war. He also emphasized that the pipeline was a major technological undertaking, requiring, in its earliest phases, 3,300 miles of pipe and over forty pumping stations. Mitterrand needed America's help in thwarting these plans, because most of the technology the Soviets required was American-made.

Reagan was paying careful attention. As the petroleum scientist Jeremy Leggett explained: "Together, the proven and undiscovered oil reserves in the former Soviet Union approach 200 billion barrels. That was almost a third of the oil ever burnt, globally, since the stuff was discovered more than a century ago. And then there is the gas. Here, the prospects are even more mouth watering for the oil companies. The former Soviet Union has more than 40 percent of the world's proven gas reserves, most of them in Siberia."[20]

Following this highly successful initial meeting, Marcel Chalet, director of the DST, secretly traveled to Washington in the sweltering humidity of late August 1981 and personally delivered over three thousand Farewell documents to George Bush Sr.—former director of the CIA and now vice president. This material included the first descriptions of the KGB's Directorate T and its leader, Leonid Sergeevich Zaitsev. In the days that followed, Weiss would learn that the Farewell documents provided minute and highly specific details concerning how science and technology were targeted, acquired, analyzed, and adapted through a concerted Soviet effort that employed over twenty thousand people. There were lists of agents in foreign countries, lists of companies, lists of friendly contacts, and lists of target contacts. Statistical records of how much technology had been targeted in previous years and evaluations of the effectiveness of KGB and GRU efforts in acquiring those targets were included. Records of what technologies had been distributed to which industry, and assessments of the strengths and weaknesses of the overall collection program, eventually allowed American, British, and French analysts to develop a complete picture of the interactions among all the Soviet bureaucracies involved in what was really a massive and ongoing case of intellectual theft.[21]

Suddenly, it was a new day. Although the United States had not yet devised a means to use this intelligence, clearly Farewell's information opened a crack into which an American wedge could be driven. If successful, this opening might eventually give the West access to Siberian energy resources in some future, post-Soviet era. The stakes were enormous, and for Weiss this news must have been accompanied by a feeling of complete vindication. His claim that technology transfer and industrial espionage had saved the Soviet Union billions of dollars in research and development spending during the cold war was now a well-established fact. He had finally earned a place at the big table.

Farewell's information convinced analysts that the Soviets had never seriously entertained détente. Weiss and other hard-line analysts brought Washington's attention back to the text of a 1972 Brezhnev briefing to the Politburo about a new phase in the cold war: "We communists have to string the capitalists along for a while. We need their credits, their agriculture, their technology. But we are going to continue massive military programs and by the mid-80s we will be in a position to return to an aggressive foreign policy designed to gain the upper hand with the West." If the Farewell documents enabled Washington to see Moscow's position on détente clearly at last, they also provided concrete proof that the Soviet economic position was precarious. In Washington the feeling grew that if America could exploit this weakness, the cold war might soon be over. In the NSC offices, the word that analysts used among themselves to describe the end of the costly contest with the Soviets is telling: it was described as the "takedown."[22]

FAREWELL

If Farewell's information was the best news to reach the West since the death of Hitler or the surrender of Japan, many still wondered where it came from. Concerning the identity and KGB function of the mole, however, very little was then known. This fact continually brought the most experienced Soviet analysts back to the question of whether or not the Farewell material could be trusted. Fortunately, many of the documents were smuggled out of the Lubyanka KBG headquarters by the mole himself, and bore the original and identifiable signatures of KGB chief Yuri Andropov and his immediate subordinates, Vladimir Kryuchov and Leonid Zaitsev, as well as those of the most elite bureaucrats known in the scientific, technological, and defense ministries of the USSR. Their provenance was impeccable. Signals intercepted with the help of America's latest Soviet defector (Victor Cheimov) supported them in every detail, and MI6's own mole, Oleg Gordievsky, who in a tidy irony had just completed an official history of the KGB for its internal use, was able to confirm some of the most incidental details, including the identity, rank, position, and job description of the mole.[23]

Although no one quite knew why he was doing it, there could be no question that Lieutenant Colonel Vladimir I. Vetrov was now working for the West. Moreover, since it was 1981, digital archives with their automatic access records had not yet become a reality. Most of the documents Farewell provided came from a filing cabinet in his own office, since he needed them in his work tracking the effectiveness of Directorate T. This made Farewell's story very similar to that of Oleg Penkovsky in the 1960s. Both men had virtually complete and unrecorded access. In the exciting days after the presentation of France's extraordinary intelligence gift to America, no one remembered that the Penkovsky affair had ended tragically with his torture and murder, and that it was followed by the disgrace, dismissal, and suicide in 1963 of his superior and drinking partner, General Serov, onetime head of the GRU.[24]

The KGB first recruited Vladimir (Volodia) Vetrov, a young Muscovite with an impeccably proletarian background, in 1959. Vetrov's biographer, the Russian journalist Sergei Kostine, identified this important period in the late 1950s as one in which "the KGB launched a massive campaign of recruiting new personnel. The Stalinist old-guard had been purged, and it was necessary to replace them. But, in addition, the Iron Curtain was lifting more and more and the Soviet secret services badly needed reinforcements."[25]

Vetrov had highly developed mathematical gifts and was also very competitive. In his teens he had been a national champion junior sprinter. Although his parents were uneducated workers, the family was fortunate to live in a privileged Moscow neighborhood, where Volodia attended one of the best grade schools in the USSR. Somehow he managed—without patronage—to get admitted in 1951 to Moscow's Baumann Technical Superior School (MVTU), the USSR'S finest engineering college. After completing a course of study that lasted over five years, the bright young Vetrov wrote his state exams in 1957 and received a posting to the SAM computer and calculating-machine factory. Although it was a secret institution, it was not a very prestigious posting. Vetrov had made lots of friends in the Soviet nomenklatura (ruling class), but they were all still young and not yet very influential.

Even so, his school friends remembered him fondly and invited him to join Dynamo, the Moscow athletic club for members of the Interior Ministry and the KGB. Dynamo welcomed this former national champion, and there he met his future wife, Svetlana, also a sprinter. On the strength of his academic record and his friends in the club, the KGB tapped Vetrov and sent him to its Dzerjinski operational school, where he again distinguished himself. Upon graduation in 1961, he was selected as a recruit to the KGB's PGU (First Chief Directorate)—an elite posting. Once again he was sent back to school. But this time it was the old "forest school" that would later become the KGB's Andropov Institute. Upon graduation, Vetrov was given a cover job in another ministry by day. At night he received intense instruction in English and French at the Lubyanka headquarters.

Vetrov had married well, he claimed respect as a former athletic champion, he was outgoing and liked, and could boast a superb education. Clearly he was being prepared for a foreign residency. During the years of his intelligence training, Svetlana had also become a national champion sprinter and had defended her title successfully twice. In 1962 she gave birth to their son, Vladik, and by 1965 she had retired from sports. The KGB sent the young couple abroad to their Paris residency, and as first postings go, it was a plum.

London, Washington, and Paris were the three most important collection centers for Directorate T. In the course of his duties in France, Vetrov came into contact with Jacques Prevost, a French scientist connected to the French industrial multinational Thomson-CSF, which had some common interests with French counter-intelligence. Prevost and Vetrov became close. A highly sociable man, Vetrov was given to Soviet-style bouts of drinking. One night, after becoming quite drunk, Vetrov totaled his residency car in an accident.

Until the 1980s, when the KGB began providing driving instruction at the Andropov Institute as part of their operational training, such accidents were common among new KGB personnel in the West.[26] Personal cars for junior officers were then a rarity in the Soviet Union, and the KGB had an internal culture of heavy drinking. It had even developed a highly effective drug (known as antipokhmelin or, colloquially, as "regional committee pills") to negate the physical effects of extreme drunkenness. Not surprisingly, automobile accidents were quite common among new residents who were unused to driving, and they were career wreckers for the young KGB agents involved, since the KGB demanded the same decorum, professionalism, and low visibility practiced by Western diplomatic personnel.

With no one else to turn to, Vetrov phoned his powerful friend Prevost, who used his influence to hush up the incident and get the car fixed overnight before it could be missed by the Soviet residency. There is some chance that the DST actually staged Vetrov's accident, since shortly afterward a DST agent introduced by Prevost approached Vetrov with a recruitment ploy. Vetrov declined the approach, and the French did not follow it with a blackmail attempt. When his posting ended, Vetrov returned to Moscow with his family as an uncompromised KGB agent. Once there, he was promoted to lieutenant colonel. This alcoholic interlude, however, was a bitter foretaste of what was to come.

In Moscow, where Vetrov worked as an analyst for Directorate T, he was popular among his colleagues but did not have a patron to support his desire to advance in the Soviet system. Despite promises of promotion by his chief of section, Vladimir Alexandrovitch Dementiev, others were continually promoted over him. For a time, he was posted to Montreal, where his superior was a nomenklatura colleague who had been his peer some years earlier in France.[27] This must have rankled the competitive and ambitious Vetrov. In any case, the two did not get along. His posting was soon canceled, and he returned to his old analyst job at Moscow Center.

By 1980 Vetrov had been a lieutenant colonel and an intelligence analyst for ten years. He had a few more years before retirement and decided to make a push to achieve the rank and paygrade of full colonel. Vetrov had considerable knowledge about the internal workings, present and past, of Directorate T, and he was aware that, despite its record of success, Directorate T had been only 37 percent effective in achieving its objectives in 1980, down from 48 percent in 1979. Moreover, Zaitsev, the chief, openly expressed his disappointment with Directorate T's performance. Gordievsky, in his history of the KGB, described one famous incident: "On one occasion in the mid-seventies . . . the head of Directorate T . . . outraged at the failure of Line X . . . officers in the United States, Western Europe and Japan to obtain a particular piece of equipment, swore violently and declared: 'In that case, I'll have to get it through our Indian contacts! I know *they* won't let me down!'"[28]

On his own initiative, Vetrov researched and wrote a report reviewing Directorate T's mid-range efficiency and made a variety of recommendations that would have improved Soviet collection abilities. His timing was impeccable, since the push to industrialize many Soviet industries (including defense) intensified in the early 1980s. But although his report attracted notice and compliments, no one was really interested in fixing something that worked moderately well enough already.[29]

This failure to achieve his desired promotion marked the beginning of a very bad period for Vetrov. Several personal disappointments, each of which would have challenged the equanimity of the strongest man, then occurred simultaneously. Svetlana, who had married him in a youthful passion against the wishes of her parents, began to have an affair with the brother of a cosmonaut, a member of the nomenklatura.[30] Vetrov knew about the affair, as did many of his colleagues, and it dealt a body blow to his self-esteem. The Vetrovs' joint campaign to secure a place for their son, Vladik, in the Economics Department at Lomonosov University, the USSR's finest school, was frustrated when the friend of a friend (a faculty member upon whose influence they had relied) refused to raise a finger for the child of a KGB officer. Without patronage, Vladik was still able to get a fairly good place at another Moscow school, but from that lower starting point he would not have the brilliant career his father hoped for. Vetrov gave up on the idea that the boy would follow him into the KGB.

Vetrov now grew quite bitter about the inequalities of the nomenklatura patronage system. The oligarchy had held him back for most of his professional career, and now it also threatened his beloved son. To his first French controller, he would soon express deep resentment toward his superiors, Dementiev and Zaitsev. Around this time, too, he began to drink heavily. In the autumn, before Vladik began his studies, and without any realistic sense of the danger in which he was placing his only child, Vetrov spoke to his son about the offer of service he had made to the French DST. Over the next eighteen months, Vetrov would tell Vladik repeatedly that it was his hope they could be exfiltrated together and go to live in the West. Although this was the dream of a desperate man, no doubt influenced by drink, it was not without some justification. Earlier in 1980 the United States had succeeded in exfiltrating Soviet cipher specialist Victor Cheimov, along with his wife and child. But in the long history of the cold war, only one other Soviet mole ever successfully reached the West from a starting point inside the Soviet Union.[31]

Unfortunately, Vetrov did not have Cheimov's good luck or his clear mind. Xavier Ameil, an intelligence amateur and Vetrov's first Moscow-based controller, must have sensed Vetrov's bitterness and his trajectory of self-destruction at their first meeting. The two men met regularly in Vetrov's blue late-model Lada, a fairly unobservable way for a Soviet citizen to speak with a Westerner in the early 1980s, when all contact with foreigners was controlled and monitored.[32] At that first meeting, the Russian engineer asked his French contact to bring something to drink for their second meeting, and thereafter every time they met Vetrov downed the better part of two bottles before dropping Ameil off and speeding home. After braving a few of these liquid monthly meetings, Ameil, who had no diplomatic protection whatsoever, was replaced by an espionage professional attached to the French Embassy.

Around this time, Vetrov also began attending regular after-hours drinking parties with his officemates. Soon he became involved in an affair with Ludmilla Otchikina, a married translator whose office was in the same corridor as his own and whom he had known for five years. During one of their assignations she found a packet of KGB documents whose significance was not lost on her. Sensing an opportunity, she began to blackmail her lover.

All his life, Vetrov seemed to desperately need constant attention and adulation. He had been a beloved only child, a star athlete, and a hero-lover to his beautiful, successful, and vivacious young wife. He had also distinguished himself again and again in the increasingly rarified atmosphere of successive KGB training centers. Now nearing fifty, his precise, mathematical mind warped by alcoholism, Vetrov knew that the brilliant promise of his young years had gone unfulfilled. He divulged details about his double life to Vladik and to Ludmilla, probably in order to win their admiration. But he also made many other mistakes that suggest he harbored a desire to be caught.

Certainly, he wanted to embarrass Dementiev and Zaitsev. Vetrov was acutely aware that—twenty years earlier—the outcome of Oleg Penkovsky's career as a mole had been the disgrace, dismissal, and suicide of the GRU's General Serov. After eighteen months of meetings with the DST, Vetrov passed nearly as many documents into Western hands as Penkovsky's impressive total. On February 22, 1982, the night before he was to hand over more documents to his French controller, Vetrov attempted to kill Ludmilla, only to succeed in killing a Moscow Militia policeman who interfered. In the weeks that followed, very little investigative effort was devoted to Ludmilla's loud and persistent claims that Vetrov was a mole inside the KGB, despite the fact that this accusation provided a clear motive for his murder attempt.

The KGB's remarkable lack of interest in Ludmilla's claims invites two possible explanations. Sympathetic or embarrassed KGB investigators may have been content to let sleeping dogs lie, since Vetrov was on his way to prison for such a long term that he would likely die there anyway. This was more or less the position of Sergei Kostine, Vetrov's biographer. A second, more likely, explanation is that the highest KGB administrators did not want to risk a Penkovsky-like scandal.[33] In the politicized atmosphere of 1982 while KGB director Yuri Andropov plotted to succeed the moribund Brezhnev, Vetrov was sentenced to serve twelve years in the Gulag for murdering a policeman. After an unnaturally long period of six months in Lefortovo Prison, he was transported to Irkutsk to serve out his sentence.

Whatever the purpose for his delay at Lefortovo, it had the salutary effect of drying Vetrov out and returning him to health before he confronted his new life in the camps. By the time he reached Camp 272/3 near Irkutsk, he had his wits about him. The challenge of prison invigorated him and—ironically—he was soon promoted to a responsible position in the internal camp hierarchy. Moreover, he began to write a series of introspective letters to Svetlana, Vladik, and his mother-in-law which show that, by the time Vetrov got to Irkutsk, he had accepted his fate and had returned to himself.

Kostine, a patriotic Russian offended by Vetrov's treason against the Soviet Union, wrote that in these letters Vetrov never expressed any regret for stabbing a policeman to death, or for betraying his country: "Any ideas of moral conscience, guilt and repentance are completely absent from his reflections."[34] But if Vetrov's resistance and his attempts to cheer his family occupy the explicit text of these letters, his guilt and regret press in from the margins. Here is a passage from one of Vetrov's earliest Gulag letters to Svetlana: "Here I am. It is impossible now to avoid punishment and regret. Sweet girl, your letter tried to console me, calm me, appeal to my reason. But I am rational now and everything is normal. I have already written to you that life here is horrible. It does no good to shake you up again . . . [and] it is impossible to describe. You must live it . . . The main curse here is hunger, that and an everyday boredom are my constant companions . . . But really, I don't know where or how to begin." By the time Vetrov reached Irkutsk—detoxified and in an environment of strictly enforced sobriety—he was confronted with the clear and immediate choice of surviving an austere KGB prison or allowing himself to sink. He simply could not afford to look back. "We live here as though it were wartime," he wrote, "and death is always present . . . The essential thing, is that I will be free someday. Everything else will then follow."[35]

Unfortunately, this was a vain hope. In 1983 the French government, unwilling to tolerate a cadre of Soviet industrial spies on French soil, expelled forty-seven Soviet citizens identified in the Farewell documents. On March 28, 1983, the Soviet official Nikolai Afanassievski was summoned to the Quay d'Orsay and shown the Directorate T documents in French possession that clearly identified these agents. According to Kostine, each officially distributed copy of the documents carried a list of names to whom the document had been distributed. The French copy included the name of Vetrov.[36]

Yves Bonnet, director of the DST at the time, specifically denied Kostine's accusation that the French accidentally exposed Farewell.[37] Nonetheless, the expulsion of the Soviet agents was a personal embarrassment to the new chairman of the Soviet Union, Yuri Andropov, who responded angrily in newspaper interviews in the Western press. If failing to prosecute Vetrov for espionage in 1982 had been an attempt to avoid embarrassing the KGB chief before he had consolidated his succession, there was no longer any reason not to do so. By 1983 Andropov had succeeded Brezhnev and also replaced the Brezhnev appointee who succeeded him—briefly—as KGB leader. Under Victor Chebrikov, Andropov's new appointee, technological disasters throughout the Soviet Union in 1983 would necessitate that the KGB find a traitor and a scapegoat. Consequently, the Vetrov case was resurrected.

Kostine put Vetrov's Moscow interrogation into the hands of Colonel Sergei Mikhailovich Golubev of the PGU's fifth directorate, the grand inquisitor of KGB counter-intelligence. He would be promoted to general by Chebrikov in the months following Vetrov's execution. However, an earlier writer—a British intelligence professional with access to American sources—identified the original interrogator dispatched to Irkutsk in 1983 to reexamine Vetrov as Colonel Vitaly Sergeevich Yurchenko, who defected to the West seven months after Vetrov's death in August 1985. According to Yurchenko, "Farewell not only promptly admitted to having spied for France but turned his admission into a blistering condemnation of the regime he had disavowed and a paean of praise for the West. He wrote out a long document entitled 'The Confession of a Traitor' which, for its well-reasoned invective, puts one in mind of Emile Zola's *J'Accuse*. His bitterest criticism was leveled against the department in which he served. The KGB's First Chief Directorate, he declared, was totally rotten, dominated by alcoholism, corruption and nepotism. His tirade ended with the words: 'My only regret is that I was not able to cause more damage to the Soviet Union and render more service to France.'"[38]

Kostine does not mention Yurchenko or Vetrov's confession, and the veracity of Yurchenko's account was called into question by subsequent events. After three months of numbing debriefings in Washington, Yurchenko one day suggested that his minders take him to lunch at a Washington restaurant near the Soviet Embassy where he had once been a KGB resident. Excusing himself briefly to use the men's room, Yurchenko literally crossed the street, redefected to the USSR, and disappeared from sight forever.

This much concerning Vetrov's final days is certain: he was returned to Moscow in late 1983 and sentenced to death for treason in the final days of 1984. During the trial he was able to speak briefly to Svetlana. They parted tenderly. On January 23, 1985, the KGB executed him more or less unnoticed in Lefortovo Prison, giving only the mildest of reprimands to Dementiev and Zaitsev. Later that year, General Golubev began his interrogation of MI6's mole, Oleg Gordievsky, with pointed questions about Vetrov. Golubev never confronted Gordievsky with any proof of treason, because he did not need to. At this point in his career, Golubev was a cagey old mongoose who knew how to sniff out a snake.

Following Vetrov's death and his own promotion, Golubev had sufficient clout to pursue Gordievsky aggressively without proof, and that is exactly what he did. Despite this formidable opponent, Gordievsky—himself a very clever man—was able to elude the KGB on their home ground. Then, to their perpetual embarrassment, he escaped to the West unaided in August 1985. That same year, Francois Mitterrand dismissed Marcel Chalet's successor, Yves Bonnet, over his handling of the Farewell affair.[39]

Although Vetrov's treason did not ruin Dementiev's career as he had hoped, it embarrassed Yuri Andropov, exposed Soviet weaknesses to the West, and had disastrous cold war consequences for the USSR.

FALLOUT

The worst short-term consequences of the Farewell case occurred in 1983 shortly before Vetrov was finally charged with treason and returned to Moscow for trial. In that year, stolen technology began to fail spectacularly in a variety of industries throughout the Soviet Union. The central figure behind these failures was the highly decorated and very odd Gus W. Weiss Jr.

It began in December 1981, around the time that the government of Poland instituted martial law and outlawed the Solidarity movement in order to appease Moscow and forestall a Czech-style invasion. The threat of Soviet aggression against Poland removed any pretense of courtesy in American–Soviet relations. Weiss, after spending several months with the Farewell material in the fall, requested a private meeting with William Casey, director of the CIA. Ordinarily, it would have been unusual for a mid-level NSC staffer to meet privately with the CIA director, but the two men had known each other for years, and the CIA had recently decorated Weiss, awarding him the Intelligence Medal of Merit for American Tradecraft Society projects from 1972 to 1980. Casey returned Weiss's call personally. When the two men met in late December, Weiss suggested a genuinely brilliant way to exploit Farewell's intelligence and undermine the stability of the USSR.

Weiss reminded Casey that the Farewell documents had included a shopping list of items that the Soviets wanted to acquire in coming years. Not only was there a shopping list, but the list rated each technological item according to its desirability and the urgency with which its acquisition was required. Farewell's own report on Directorate T's efficiency noted that the most urgently desired A-list items were also the most quickly and completely acquired. In other words, American intelligence had more or less certain knowledge about what the Soviets needed to steal and what they would try to steal next. Moreover, Farewell had also provided a full list of Line X personnel as well as the Western companies they had already infiltrated and those they had targeted for infiltration. Finding Line X agents then at work in the United States would be a very easy task.

So "what if," Weiss wondered, "we applied American know-how to these A-list items and doctored them, so that they would appear like the genuine article upon their acquisition by the Soviets, but would later fizzle."[40] The KGB would have put an enormous amount of time, money, and effort into acquiring products that would turn out to be very short-lived. This would make Directorate T look bad indeed, Weiss said, and it would also destroy the confidence of Soviet industrialists in illegally acquired technology. Furthermore, if the American ruse was uncovered, Weiss continued, the results would actually be worse for the Soviets, since they would not know which stolen technology to trust and which to discard.

Throughout the meeting, Casey listened silently and appreciatively. He was about to leave, when Weiss pitched his final ball. "We don't actually have to leave it at planned obsolescence," he told Casey. "We could take it a step further. We *could* shake their morale completely by making some of the stolen items malfunction in spectacular ways." At this, the CIA director settled back thoughtfully. "I'm thinking of their most crucial areas," Weiss said. "You know, defense and the pipeline." After he had heard all Weiss had to say, Casey made a highly appreciative but carefully noncommittal noise in his throat.

Then, in the early days of the new year, after reviewing his personal record of the meeting, Casey took Weiss's ideas privately to President Reagan and received enthusiastic approval. Later, in a general executive meeting outlining a range of anti-Soviet strategies, President Reagan also committed himself to the MIRV and the Strategic Defense Initiative (SDI, or "Star Wars") and to a general policy of investment in advanced military technologies.[41]

What would follow was a collaborative counter-intelligence effort in the best tradition of the American Tradecraft Society. The CIA, FBI, and the Department of Defense quickly identified Soviet agents engaged in acquisitions projects and won the cooperation of American industry in sabotaging the products before they were acquired. When the president of Texas Instruments was tapped for assistance, he "allowed one of his company's chip-testing devices to be made available for Soviet interception in Rotterdam. The machine was modified to work initially as expected, but after a few trust-winning months, it would salt its output with defective chips . . . Only later did it register [with the Soviets] that some of the chips might end up in the Soviet Strategic Missile program."[42]

Planned obsolescence and even malfunction was a small but

vital piece of the overall strategy of economic warfare against the Soviet Union that was being put in place by Casey and by Caspar Weinberger at the Defense Department. Another part of the emerging strategy was to limit the Soviet Union's access to Western credit. Roger Robinson, a New York investment banker, had discovered that the Soviets were actually doubling the loans they had taken out on the future success of the Urengoi and Sakhalin Island pipelines. Everything was ridiculously overfinanced, and the USSR was taking a massive gamble with enormous sums of borrowed money. By shutting off credit, the Reagan administration delayed pipeline development and simultaneously raised Soviet costs, forcing them to exhaust their limited reserves of hard foreign currency. The unpopularity of the USSR's role in the military clampdown in Poland made a worldwide credit embargo feasible. With President Reagan's blessing, Robinson, now an NSC staffer, began a series of European meetings aimed at denying further credit to the overextended Soviets. His secret presentations to European bankers successfully demonstrated the strained financing and poor risk of the Soviet pipeline project.[43] The USSR soon found itself cut off from foreign credit.

The Soviets had reached a critical juncture in the pipeline's development and had already contracted for supplies of pipe, turbines, drilling equipment, and computer software from many American manufacturers. When the U.S. government acted to prevent the export of these materials, the Soviets grew desperate and tasked the thirty members of their Cologne-based delegation with finding the suppliers they needed.[44] With America's urging, Alsthom-Atlantique, the Soviet's French supplier of turbines, refused to sell across the Iron Curtain. In desperation, the Soviets pulled their brightest people off existing projects and dipped into their hard currency reserves to manufacture rotor shafts and blades. A Soviet engineer who worked on the project recalled: "We tried to build a twenty-five-megawatt turbine, [and] threw resources into a crash program. [But] We failed. It was a huge drain on our resources. It cost us dearly."[45] Similar efforts to create a sheet steel factory that would have enabled the USSR to manufacture its own high-pressure piping met with enormous costs and delays.

But the crucial software necessary to control the pipeline could not be homegrown or reverse engineered. As Derek Leebaert reported in 2002, when "the Soviets contrived . . . to obtain it illegally through Canada, the FBI doctored the coding to cause destructive power overload once applied in Russia, inducing surges and causing sundry destruction. At least one immense (and deadly) pipeline explosion could be seen from space."[46] Leebaert's account attracted little notice, and Gus Weiss was ruled to have committed suicide in November 2003, one year after it was published and shortly before the invasion of Iraq. Although Weiss had experienced depression during his life and had been treated for alcoholism in the 1980s, his hairlessness and extreme body modesty make the method of his suicide remarkable. At age 73, this highly decorated American is said to have jumped, scantily clad, from the balcony of his Watergate apartment, leaving the bedroom door locked behind him. Not only did his death silence an outspoken and especially well-informed critic of the Iraqi invasion, it also prevented any complete first-hand account of the pipeline explosion from being passed into history. News of the explosion would have remained obscure had not William Safire reviewed and excerpted a second account of Weiss's masterstroke by a former secretary of the Air Force, Thomas C. Reed.[47]

In 2004 Reed published *At the Abyss: An Insider's History of the Cold War*. This account, based on personal memories of his White House years, drew sharp and immediate criticism in the *Moscow*

News from former KGB personnel. Reed has since acknowledged that he may have had the exact date of the explosion wrong, but it occurred sometime in 1983. Here is how he described the Urengoi 6 disaster:

> [In] the Weiss project . . . pseudo-software disrupted factory output. Flawed but convincing ideas on stealth, attack aircraft and space defense made their way into Soviet ministries . . . The Soviets [also] needed sophisticated control systems . . . [but] when the Russian pipeline authorities approached the U.S. for the necessary software, they were turned down. Undaunted the Soviets looked elsewhere: a KGB operative was sent to penetrate a Canadian software supplier in an attempt to steal the needed codes. U.S. intelligence . . . in cooperation with some outraged Canadians "improved" the software before sending it on . . . Buried in the stolen Canadian goods . . . was a Trojan Horse . . . The pipeline software that was to run the pumps, turbines and valves was programmed to go haywire . . . To reset pump speeds and valve settings and to produce pressures far beyond those acceptable to the pipeline joints and welds.[48]

A three-kiloton blast, "the most monumental non-nuclear explosion and fire ever seen from space," puzzled White House staffers and NATO analysts until "Gus Weiss came down the hall to tell his fellow NSC staffers not to worry." Two full years before the NATO allies rolled up the KGB's international Line X networks, the Soviets sustained significant economic damage from Weiss's efforts at sabotage.[49]

Jeremy Leggett, who visited Siberia in 1991, has described the results of the inexpert and badly funded Soviet effort to develop energy resources during the period following America's targeted sabotage of Siberian pipeline projects: "Significant spills and leaks were commonplace . . . The loss of oil in the early 1980s amounted to 1.5 percent of the total extracted in the area . . . [Each year] 80 times the amount of oil spilt during the Exxon *Valdes* disaster was simply being shed into the soils, rivers, and lakes of the region. In the Tyumen area, when the snows melt in summer, vast standing lakes of oil can be seen. The biggest at Samotlor, was 11 kilometers long and 1.5 to 2 meters deep."[50]

The 1983 Urengoi 6 explosion would have been sufficient reason for the KGB to exhume Vetrov from the living death of Gulag 272/3–Irkutsk and make him the scapegoat of record shortly before executing him. But 1983 saw a much more frightening technological failure than that of the Siberian gas pipeline. This second failure sent chills through the coolest hearts of the USSR's cold warriors and guaranteed Vetrov's execution.

In June 1983, only three months after President Reagan publicly announced SDI, which Yuri Andropov immediately condemned as "insane," a Soviet satellite equipped with stolen American computer chips suddenly went wildly defective. This raised the Soviet military's anti-American paranoia to an all-time high. Defects in Soviet satellites had happened before, but in this case the satellite in question was "the Soviets' only reliable means of detecting a U.S. [missile] launch." The satellite "registered [American] missiles pouring out of the silos. Solely because the duty officer of the day came from the algorithm department . . . [did the Soviets] sense that the alert was inauthentic."[51]

Here then was the full measure of Gus Weiss's victory over the illegal transfer of American technology to the Soviet Union. By 1983 the USSR's plans to finance their own resurrection had been left in tatters by the stalled Urengoi pipeline project. Since 1981, when American economic warfare against the USSR began in earnest, the Soviet Union had experienced losses totaling $3 billion per year in foreign trade, an area that had experienced a $270 million surplus in 1980. Gus Weiss's sabotage had left them powerless to create their own pipeline infrastructure or to repair the one they had largely stolen from the West. Moreover, the Soviets were deeply in debt to their Western creditors and now also completely vulnerable to attack, since they could no longer trust the early warning systems that had also been created by and stolen from Western companies. This excruciating moment of vulnerability occurred in the summer of 1983, when, according to Weiss's immediate NSC superior, "the two blocs were closer to hot war than at any time since the 1962 missile crisis."[52]

Fortunately, the hardliner Yuri Andropov would eventually be replaced by Mikhail Gorbachev. Within the decade, glasnost and perestroika would follow—partly, at least, because of the intelligence roles played by an American patriot and a Soviet traitor. Very soon after Vetrov's death, Gorbachev would trade an interest in the development of Siberian energy resources to Western companies. Ironically, by the end of 1991 more than thirty-six oil companies had set up headquarters in Moscow.[53]

Today, America's sabotage of energy development in Siberia and the subsequent privatization of these energy resources are once again highly politically charged issues. Within the civil service of the Russian Republic, a new nomenklatura, the siloviki (comprised mainly of former KGB officers), is desperate to regain control of its own resources from privatized companies like YUKOS or from organizations like BP-TNK, which is 50 percent foreign-owned. For this reason, Rosneft, the Russian national energy giant, has stripped YUKOS of its most valuable holdings and prevented BP-TNK from bidding on new natural resource exploration licenses.

For Vladimir Putin and the siloviki, the quest to control and exploit Siberian oil and gas under a government umbrella represents exactly the same promise it had for the Soviets in Gus Weiss's day. In an era of soaring energy prices, Siberian energy development is a potential cash cow that could finance a new era of badly needed Russian prosperity by developing the easternmost Siberian oil fields for strategically sensitive sales to North Korea, China, and Japan. Realistically speaking, it is still not to the advantage of the siloviki or the SVR (successor to the KGB) to confirm the success of the Weiss project's sabotage. Similarly, it is not to the advantage of Western oil companies to allow Putin's nationalistic oil and gas ambitions to succeed.

Like most problems with technology, pollution is a problem of scale. The biosphere might have been able to tolerate our dirty old friends coal and oil if we'd burned them gradually. But how long can it withstand a blaze of consumption so frenzied that the dark side of this planet glows like a fanned ember in the night of space?

RONALD WRIGHT, *A SHORT HISTORY OF PROGRESS* (2003)

9   Cell Phones and E-Waste

Electronic components have extremely short lives. In the United States, cell phones built to last five years are now retired after only eighteen months of use. These and other ubiquitous products, like televisions, which are owned by more than 90 percent of the population, are creating unmanageable mounds of electronic waste each time they are thrown away. All of the discarded components in this growing mountain of e-waste contain high levels of permanent biological toxins (PBTs), ranging from arsenic, antimony, beryllium, and cadmium to lead, nickel, and zinc. When e-waste is burned anywhere in the world, dioxins, furans, and other pollutants are released into the air, with potentially disastrous health consequences around the globe. When e-waste is buried in a landfill, PBTs eventually seep into the groundwater, poisoning it.[1]

As enforcement of the United Nations' Basel Convention on the Control of Trans-Boundary Movements of Hazardous Wastes and Their Disposal becomes stricter around the world, the United States will soon be prevented from exporting its discarded electronic products to developing countries for burning, burial, or dangerously unregulated disassembly. As the waste piles up in the United States, above and below ground, contamination of America's fresh water supply from e-waste may soon become the greatest biohazard facing the entire continent.

In 2001 the Silicon Valley Toxics Coalition estimated that the amount of electronic consumer waste entering America's landfills that year would be between 5 and 7 million tons.[2] This represented a substantial increase over the 1.8 million tons of e-waste produced in 1999, the first year that the EPA tracked hazardous waste from electronic products.[3] But we have seen nothing yet. By 2009, the overall amount of American e-waste will jump radically, when the FCC-mandated shift to high-definition television goes into effect—a one-time instance of planned obsolescence with unprecedented negative consequences.

Because the toxins contained in most electronics are indestructible, the European Union has banned their use by manufacturers and consumers. This ban is proving to be an effective encouragement to the development of alternative, non-toxic materials for electronic manufacture. In the United States, by contrast, the EPA-chartered organization NEPSI (National Electronic Product Stewardship Initiative) failed, in the early months of 2004, to gain support for federal-level legislation that would make American manufacturers financially responsible for reclaiming the toxic components of e-waste. Although some legislation now exists at the state level, there is no uniformity, no consistency, and no funding for electronic waste disposal programs throughout the United States.

The increasingly short life span of high-volume electronic goods, along with miniaturization, is what causes the e-waste problem. This lack of durability, in turn, grows from a unique combination of psychological and technological obsolescence. As we have seen in the preceding chapters, psychological obsolescence is one of a complex set of corporate strategies first adopted in the 1920s to confront the challenge of overproduction by creating an endlessly renewable market for goods. Much earlier—in 1832—the Cambridge mathematics professor Charles Babbage first described technological obsolescence as an inherent (but unnamed) phenomenon created by the industrial revolution.[4] Both kinds of obsolescence have since been wholeheartedly embraced by American industry, for all kinds of goods, from pantyhose to skateboards. But as Moore's Law predicts, the rate at which consumer electronics become obsolete is unique, because of integrated chip technology and miniaturization.

At first glance, it might seem that responsibility for the looming crisis of electronic waste can simply be laid at the door of greedy IT manufacturers and marketers. But identifying who actually creates e-waste is intricate. As we will see in this final chapter, the actions and habits of America's consumers threaten to flood the world with toxins, just as surely as do the misguided priorities of multinational corporations. Now more than ever, end-users of new technology need to pursue higher levels of technological literacy in order to negotiate the complex interactions among technology, society, and the environment. Ignorance of these interactions effectively grants a permission slip for technological hazards to persist.

WHY DOES E-WASTE OCCUR?

By 2002 over 130 million still-working portable phones were retired in the United States. Cell phones have now achieved the dubious distinction of having the shortest life cycle of any electronic consumer product in the country, and their life span is still declining. In Japan, they are discarded within a year of purchase. Cell phones have become the avant-garde of a fast-growing trend toward throwaway electronic products. Not only are rates of cell phone replacement rapidly increasing, but previous estimates of maximum market penetration are proving inaccurate. In 2000, marketing experts predicted that, at most, 75 to 80 percent of people in industrialized countries would own a cell phone in the near future. But, as one study has since pointed out, "in some countries . . . such as Japan, Finland, and Norway, penetration will exceed 100 percent within the next few years," and "within a few years having just one cell phone will seem as odd to most people as owning a single pair of shoes."[5]

In other words, cell phone e-waste is growing exponentially because people who already have cell phones are replacing them with newer models, people who do not have cell phones already are getting their first ones (which they too will replace within approximately eighteen months), and, at least in some parts of the world, people who have only one cell phone are getting a second or third. Such a pattern renders the term "obsolescence" itself obsolete. It makes no sense to call a discarded but working phone obsolete when the same make and model is still available for purchase and continues to provide excellent service to its owners. In 2005 about 50,000 tons of these so-called obsolete phones were "retired," and only a fraction of them were disassembled for reuse. Altogether, about 250,000 tons of discarded but still usable cell phones sit in stockpiles in America, awaiting dismantling or disposal.[6] We are standing on the precipice of an insurmountable e-waste storage problem that no landfill program so far imagined will be able to solve.

How does e-waste happen? As I have tried to document in the preceding chapters, modern consumers tend to value whatever is new and original over what is old, traditional, durable, or used. Advertising and other marketing strategies have helped create this preference by encouraging dissatisfaction with the material goods we already have, and emphasizing the allure of goods we do not yet own. When dissatisfaction and desire reach a peak, we acquire the new and discard the old. Electronic waste is simply the most extreme version of this consumer behavior. In the words of the Silicon Valley Toxics Coalition: "Where once consumers purchased a stereo console or television set with the expectation that it would last for a decade or more, the increasingly rapid evolution of technology has effectively rendered everything 'disposable.'"[7]

Contemporary critics generally blame advertisers alone for the perpetual dissatisfaction that fuels our throwaway culture. The average American, one analyst has noted, will have watched more than three years of television advertising by the end of his or her life. Despite this shocking amount of lifetime exposure to TV ads, blaming the rapacity of Americans' consumption entirely on manipulative advertising is simplistic. Although Vance Packard's conspiracy theory alarmed America when it first appeared in *The Hidden Persuaders,* forty years later his explanation is demonstrably incomplete. In particular, it ignores what one sociologist of consumerism, Colin Campbell, has described as the "mystery" of modern consumption itself—"its character as an activity which involves an apparently endless pursuit of wants, the most characteristic feature of modern consumption being this insatiability."[8]

Campbell's analysis of the mechanics of consumerism is one of a handful of inquiries that look at product demand from the consumer's perspective. He acknowledges that manufacturers often produce goods that become quickly unusable and that the "definitional threshold of what 'worn out' might mean" is a moving target. But Campbell refuses to blame manufacturers or marketers alone for the massive amounts of waste produced by our throwaway culture. Instead, his study tries to make sense of the fascinating attitudes, beliefs, and behaviors that surround the acquisition of new goods. Campbell even examines the word "new" itself, to discern the epistemological foundations of what he calls our "neophilia," or love of new things. His insights on consumers' motivations are especially significant to an understanding of the global rush to purchase new technologies like cell phones, and to the subsequent problem of e-waste produced by their rapid consumption.[9]

According to Campbell, neophiliacs come in three varieties. The first kind acquires new products and discards older ones in order to sustain a "pristine" self-image. These pristinians, Campbell writes, are obsessed with whatever is fresh or untouched. They want to live in new houses, drive new cars, and wear new clothing. They immediately replace anything—from furniture to plumbing fixtures—that bears the slightest sign of wear. Why they do this is not entirely clear. Sometimes this behavior may be an overreaction to a personal history of poverty or emotional deprivation. But some sociologists believe it may simply derive from a pressing need to assert and validate one's recently acquired social status.[10]

Campbell himself is very cautious in drawing any general conclusions about this group of pristinians, other than to note that while these individuals are avid consumers, their tastes are markedly conservative. They are largely indifferent to changes in style, often buying new items that are identical to the ones they replace.[11] Compared with the other two groups, pristinians are the most resistant to innovative technologies.

Much more impressionable are Campbell's second and third

kinds of neophiliac. He describes the second type as "trailblazing consumers," people who crave the newest product lines and the very latest technology. This kind of consumption is found most commonly among technophiles. Such early adopters tend to make very good consumers, Campbell notes, because they are the first to recognize a useful new device. These knowledgeable techies play a key role in leading others to accept new technologies.[12]

The vast majority of neophiliacs fit neatly into Campbell's third category, which we might call the fashion fanatics. These people are hypersensitive to the latest styles, and this sensitivity "creates a rapidly changing and continuous sequence of new wants." Such individuals, Campbell writes, are highly stimulated by new and original products, and they respond with boredom to whatever is familiar. This largest group of neophiliacs are fickle consumers who change their product preferences continuously and quickly, a fact that is especially obvious in their styles of dress. In addition, their sensitivity to fashion causes a very high rate of "want turnover" in cultural product areas like records, films, and books. They respond with enthusiasm to almost any retail novelty that offers a new experience or sensation. Although these "lovers of the exotic" do not fit neatly into any traditional demographic, Campbell notes that fashion sensitivity is rare among the old and much more common among adolescents and young adults, and that, in every age group, women are much more fashion-conscious than men.[13]

A recent exploration of conformity and dissent includes a description of a sociological phenomenon that seems relevant to neophilia: the "social cascade." It starts when "one or a few people engage in certain acts" and then other people soon follow these leaders, either in order to be right or simply to gain social approval. "Influenced by the decisions of their increasingly numerous predecessors," a majority of the population eventually commit to the new behavior. Such cascade effects explain the sudden popularity of restaurants, toys, books, movies, and clothing. They are also especially common among IT products like cell phones.[14]

Combining the notion of social cascades with Campbell's categories of neophiliacs, we can begin to chart a trajectory of adoption as the cascade effect moves through different types of IT consumers. The first to use a new electronic product is a small group of technophiles. These trailblazers pass their interest and enthusiasm on to a more fashion-conscious mass of predominantly young people. Eventually, when the trend to adopt the new product becomes exponential, as it did with the acceptance of the Internet, for example, even the resistant pristinians consider adopting the new technology to some extent.

Campbell points out that IT products are especially susceptible to such cascades because these "products themselves . . . [can] serve as an important channel of information about recent developments."[15] In other words, IT products create a sort of loop effect, by passing along information to their purchasers that hastens their own replacement. These effects are most visible in products like iPods, PS2s, and PSPs. The spectacular accumulation of electronic waste results from both the initial sales cascade and the heightened and repetitive obsolescence of IT technology. Invariably, after you buy the newest electronic widget, you dump the old one.

Today, America is participating in a worldwide cascade of acquiring and reacquiring cellular telephones. The sociologist Rich Ling offers this global perspective: "On a worldwide basis, there is, roughly speaking, one mobile telephone subscription for every fifth or sixth person . . . there are slightly more mobile telephone subscriptions than traditional landline subscriptions . . . Less than one-quarter are in the Americas." Among the billion or so cell phone users around the world, the vast majority belong to Campbell's third neophiliac category. From Hong Kong to Amsterdam, the most dedicated users of cell phones are fashion-conscious young people. Older people prefer the more familiar "Plain Old Telephone Service," and are less like to rely on cell phones when POTS is available. Ling's explanation of the worldwide popularity of mobile phones among the young vividly recalls the important role that conformity plays in lives of adolescents. Adolescence is a socially definitive period when people work at developing both an identity and self-esteem. In Ling's view, their "adoption of the mobile telephone is not simply the action of an individual but, rather, of individuals aligning themselves with the peer culture in which they participate . . . The cover, the type, and the functions are a symbolic form of communication. These dimensions indicate something about the owner . . . The very ownership of a mobile telephone indicates that the owner is socially connected."[16]

Ling notes that an adolescent's peer group provides him or her with "self-esteem, reciprocal self-disclosure, emotional support, advice, and information." For the first time in their lives, young people are able to seek a perspective on social interactions from a source outside their family group. They communicate with one another in order to learn "how to be" within a closed circle of equals. It comes as no surprise, then, that modern adolescents use cell phones to establish (and police) a tight, extra-familial community in which—paradoxically—they nurture their independence while drawing a symbolic boundary around themselves that resists intrusions. Such "idiocultures" often include systems of nicknames, as well as consumable items like clothing, music, and other accessories. Adolescence, Ling reminds us, "is that period when the peer group and friends are most central. We see this in the unquenchable desire of teens to be together with and to communicate with their friends . . . Here we can see where mobile telephony fits into the picture."[17]

In group interviews with young adults—all cell phone users—Ling raised the issue of repetitive consumption. The comments of these young men and women confirm Campbell's suggestions about the importance of fashion and product-boredom as driving forces among the largest group of fashion-conscious consumers, the young. In vivid detail, these subjects described how group membership depends on quick-moving trends in color, model, ring tone, screen logo, and messaging lingo. The richest feature of Ling's fascinating book is his focus group studies of these former adolescents:

> MARTIN (23): The mobile phone is like clothes . . .
>
> CARLOS (25): If you have a Nokia you are cool; if you have a Motorola or a Sony-Ericsson you're a business guy . . .
>
> ANDERS (22): If you don't have a mobile, you are out of it! The model has a lot to say, you know. A Phillips "Fizz" from 1995 is nothing that you show off.
>
> HAROLD (24): I think that blocks of cement [slang for older Phillips phones] are cool.
>
> CARLOS (25): I am proud of my Motorola Timeport.
>
> PETER (24): If you have a Nokia you are one of the herd [a positive connotation]; if you have something else, you will soon buy a Nokia.[18]

Cell phone ownership has not yet reached a rate among American teens that can compare with rates in Norway, where Ling, an American sociologist, currently conducts his research. Adolescents in such "mobile-intense" societies as Italy, Japan, the Philippines, Scandinavia, and South Korea possess not one but two phones at any given time. This is not the case in the United States. A partial explanation for the difference is that in the international system of billing, the "calling party pays." In the American system, both parties share the cost, whether they place the call or not. Extensive cell phone use can be expensive for American teens, and consequently the highest adoption rates for cell phones in the United States are still among mature adults, although this is quickly changing.[19]

Economic considerations also explain the popularity of SMS (Short Message System) text messages in Norway and other European countries. Until 2002 American teens used instant messaging (IM) programs over Internet terminals (PCs), but they did not rely on phone messaging programs, which are cumbersome and more expensive in the United States than elsewhere. The peculiarities of America's billing system and the lack of a standardized American texting protocol combined to make cell phone messaging more expensive and less user-friendly in the United States than in the rest of the world. But after the summer of 2002, when SMS was adopted in America, the number of text messages tripled in a two-year period. As John Grisham's thriller, *The Broker*, made clear, text messaging quickly became very popular: by the middle of 2004, Americans—mainly adolescents—were sending roughly 2.5 billion text messages a month, despite the fact that at about ten cents a message for both the sender and receiver, texting is still very expensive for American teens. This latest demand for text messaging capability is fanning the flames of cell phone obsolescence among teenagers.[20]

Text messaging on cell phones is yet another example of the cascade effect among neophiliacs. In Europe, the phenomenon has already grown beyond the initial market of trailblazing technophiles to include younger, fashion-addicted neophiliacs. In the United States, text messaging lagged until the SMS protocol was adopted, but afterward it quickly began to catch up. By 2005, American teens' observations about how text messaging contributes to their culture exactly repeated those of Norwegian teenagers: "'It's about feeling part of a little group with cell phones,' Denise said. 'You want to learn what's going on.' Karina agreed. 'It's about belonging,' she said."[21]

The smaller cell phone market share represented by America's adult users falls into Campbell's second category. To some degree, adult users are technophiles accustomed to successive upgrades and replacements for consumer electronic goods. They are less interested in the self-display and self-identity issues of adolescent users. What is most important to this group is the ability of the product to meet their needs. Such people discard and consume products like cell phones repetitively not for reasons of fashion but because as consumers they take it for granted that science and technology together will "produce a continuous flow of inventions and improved products."[22]

MICROCOORDINATION

Perhaps because his grandfather was an electrical engineer for Mountain Bell, Ling is especially attuned to trailblazing cell phone digerati. Clearly, he is an advocate of mobile technology, though he is careful to place its adoption in an appropriate historical context in order to show how our responses to this phenomenon emerged culturally. Ling sees the development and proliferation of the cell phone as an extension of a series of inventions that includes railways, standard time, the telephone, the automobile, and the personal timepiece (both pocket watch and wristwatch). What these innovations have in common is their ability to coordinate human social interactions. The advantage and attraction of the cell phone is that it permits a new micro-level of social coordination previously unavailable and indeed unimaginable.

Without question, safety and security factor into the initial decision to own a cell phone, as Ling points out. Cell phone purchases in the United States went sky high right after the 9/11 attacks on New York and Washington. But the main function of cell phones is not to provide a sense of safety and security, according to Ling. It is, rather, to permit a widespread "softening of schedules" that would be impossible with wire-based POTS. Traditional landline telephones are used primarily to carry out coordinating activities such as making appointments and organizing one's schedule. The overwhelming advantage of the cell phone, Ling asserts, is that it extends these possibilities almost to the vanishing point, allowing us to replan activities "any time and anywhere—to a greater degree than . . . the traditional landline telephone" allows.[23]

In a postmodern world that relies on intricately connected transportation systems, the need for geographically dispersed coordination among small groups of people is acute. Beginning with the automobile, transportation systems have been the most significant innovations in the evolution of modern cities, which have changed from a series of concentric rings to a complex integration of highways, shopping centers, and strip development. Ling writes that as a result of this geographic complexity, effective coordination has become critical. The high mobility of our deadline-based society, with its transitory habits of work, casts in stark relief the inflexibility of a wired system of telephone communication. Cellular technology, by contrast, obviates the tyranny of fixed times for meetings, because "meetings can be renegotiated and redirected in real time . . . With access to mobile communication, we can quickly call to see if our meeting partner will make the date."[24] Partners to a meeting no longer have to be office-bound or geographically located in order to communicate about schedules. They can "renegotiate their plans . . . 'on the fly.'" This flexibility marks a movement away from a belief in "a type of linear . . . time, in which meetings, social engagements . . . are fixed points," and toward a mental space in which everything is in continual negotiation. Clearly this is a McLuhanesque shift in which a new medium can be seen shaping new consciousness.[25]

Ling expands his microcoordination hypothesis by putting it into the context of a series of accepted inventions that have transformed or retransformed our notions of social space and time. Ling's most fundamental claim is that we are now at just such a historic juncture with cell phones, as we struggle to work out the terms of our acceptance of these seemingly unbounded communication tools. Our forefathers once did the same with their pocket watches and wristwatches, and with telegraphy, early telephones, and automobiles. The device is still new "and has not yet found its natural place."[26]

One of Ling's most interesting observations concerns the ambiguous role of timekeeping in relation to microcoordination. Although it is too early to know whether mechanical timekeeping is in competition with, or simply supplemented by, mobile technology, it is clear that wristwatches for the purpose of timekeeping are no longer essential consumer items: "Extremely precise clocks have been included in everything from microwave ovens to video players. They are so widespread that even if your own watch does not function . . . there are myriad alternatives freely available."[27]

In his comparison between wristwatches and cell phones, Ling notes that wristwatch–based timekeeping is the more mature technology, one that has found "a stable and taken-for-granted place on the body," whereas the mobile telephone "is not yet taken for granted, nor has it found its final locus on the body." Today, various possibilities for such a locus are being explored. Korean students, for example, now wear their mobiles on a neckstrap. In North America, hip and belt holsters have emerged, though they have not caught on in many demographic groups. A variety of recommendations and prototypes for wearable devices have also emerged. These efforts to find an appropriate, comfortable, and safe body location for cell phones confirm Ling's observation that phones are currently in a transitional phase of social acceptance and adoption.[28]

As mobile technology settles into place, wristwatch manufacturers are making a huge—if largely unnoticed—effort to find new possibilities for extending their line. Casio, for example, has launched a development effort that has produced many new wristwatch functions in anticipation of the impending obsolescence of watches as timekeeping devices. Casio's "Wrist Technology Family" combines digital timekeeping with MP3 players, cameras, voice recorders, GPS receivers, and personal organizers. Unfortunately, as David Pogue, a technology columnist for the *New York Times,* points out, due to the miniaturization of these devices, "thoughtful ergonomics are the first to go." His description of the truly useful ProTek Satellite Navi Watch with built-in GPS receiver highlights a difficulty of the new tools: "It's like wearing a minivan on your wrist." In addition to Casio's new uses for the wristwatch, various wrist-worn GPS locater devices had been designed to "facilitate care for our children, the chronically ill and the elderly, including Wherify and Digital Angel." The marketing associated with such experiments focuses on how these devices participate in the trend toward convergent technologies.[29]

But, in truth, the rush by wristwatch manufacturers to find new devices to wear on our wrists is driven by the knowledge that single-function timekeeping wristwatches are obsolete. And all of the convergent wrist devices designed to replace them are still missing some desirable functionality or quality. Most often, what's missing is simply sensible, user-friendly ergonomic design, but in some cases it is common sense. For example, how many people will be attracted to a digital watch that doubles as a cigarette lighter? This kind of "convergence" has an air of desperation about it. In the mobile world our children inhabit, wristwatches have joined the company of pay phones and CRTs, as products of another time and generation.

WHAT CAN BE DONE ABOUT E-WASTE?

Although the CRTs of PC monitors and analog TVs contain the highest concentrations of toxins among the different varieties of e-waste, it is, ironically, the small size of cell phones that makes them a significant toxic hazard. Disassembling tiny components in order to recover their parts and materials for reuse is expensive. Also, because they are so small, most phones are simply tossed into the trash, and from there they travel to incinerators and landfills. The number of these discarded miniaturized devices now threatens to "exceed that of wired, brown goods." But is there really any alternative?[30]

In *The Green Imperative*, Victor J. Papanek pointed out that by turning down the quick buck to be made from blindly following the miniaturizing trend, manufacturers need not lose financially.

They can simply charge a bit more for more durable, better-designed goods that are more amenable to disassembly and reuse. In contemporary American engineering circles, topics inspired by this aesthetic of green design are beginning to dominate meetings of the Institute for Electrical and Electronics Engineering (IEEE). The popularity of papers addressing Life Cycle Assessments (LCAs) and Extended Product Responsibility (EPR) may indicate that the electronics industry is now undergoing positive change from within. Other indications of this change are the cost-free take-back, reuse, and recycling programs now in place at most major American electronics manufacturers, including Hewlett Packard and Dell. Despite these changes, however, the cultural expectations of both industrial designers and consumers must still undergo a radical shift in order to accommodate the complexities of twenty-first-century life.

E-waste is not the only problem associated with cell phones. The introduction of every new technology brings with it a complex set of challenges, some of which cannot be recognized initially. In the 1980s, these were referred to as "technological hazards." Today, they are usually described with the more neutral term "technological risk," a phrase which encompasses many more issues than the creation and disposal of hazardous e-waste. While pioneering methods to facilitate our ability to recognize, predict, and confront such risks have proliferated in the scientific and technological literature, this knowledge has not percolated into the public's awareness. Increasing technological risk needs to be balanced with an increasingly technologically literate populace. Although this will require extensive and fundamental changes in American education, the stakes are enormous.[31]

All of the current questions raised about cell phone manufacture and use—ranging from traffic accidents to possible carcinogenic effects to the mining hazards associated with obtaining requisite minerals—point to the importance of developing and maintaining technological literacy. Of course, keeping their customers informed about the risks as well as the benefits of technology may not always be in the corporate best interest, when measured solely by the bottom line. In 2001 George Carlo and Martin Schram wrote a useful but little-known book about the Cellular Telecommunications Industry Association's (CTIA's) efforts to discredit the handful of independent scientific studies that have linked extensive early cell phone use to traffic injuries and possibly brain cancer.[32] Today, technologically unaware consumers dismiss these issues as alarmist hoaxes, and disinterested research into these questions remains underfunded.

The need for technological literacy is especially visible in the case of mining risks. Almost no one foresaw that the worldwide demand for colombo-tantalum ore, or coltan, would one day skyrocket in the cell phone industry, or that it would produce social and geopolitical chaos in West Africa, where the ore is most plentiful. Coltan is essential to the manufacture of cell phones and other electronic products because it can be refined to produce tantalum, a critical ingredient in capacitors. Tantalum capacitors are used in almost every cell phone, pager, organizer, and laptop.[33] The fact that only a fraction of the most informed cell phone consumers are aware that coltan mining produces economic devastation points out the two-sided nature of technological literacy. Consumers must be alerted to the importance of technological reportage, and trained to understand it. But news organizations must also make important technological stories more accessible to the public. Only a public that tries to understand the consequences of coltan mining can begin to make an informed choice about the global trade-offs associated with "trading up" to a new and better cell phone. As technology makes the world smaller and more intricately connected, the issue of technological literacy becomes increasingly vital.

Another global problem associated with e-waste has to do with "recycling" in developing countries. "Recycling" is a word with unexamined positive connotations for most Americans, but "recycling" can obscure a host of ills. Until very recently, e-waste has moved surreptitiously and illegally from North American stockpiles to countries in the developing world. Once there, a complex network of cottage industries salvaged what it could from America's discarded electronics, in a variety of unregulated and unsafe ways. In the year 2000, *Exporting Harm,* a report by the Seattle-based Basel Action Network, co-authored with the Silicon Valley Toxics Coalition, made the Guiyu region of southeast China famous for a piecework computer recycling operation conducted without protection either to the local workers or to their environment.[34] As a result of the bad publicity generated by this single report, e-waste salvage efforts have all but ceased in Guiyu, if not in China itself.

But Guiyu is only the best known of an array of similar operations throughout Asia. In India, Pakistan, and Bangladesh, unregulated facilities burn excess plastic waste around the clock, pumping PBDE and dioxin-laden fumes into the air. Despite respiratory disorders and skin diseases among local residents, and despite transoceanic airborne contamination, these facilities are still considered valuable local businesses. Countries facing economic desperation welcome the hard currency circulated by quick and dirty reclamation of the heavy metals and chemicals contained in America's electronic waste.

To date, the United States is one of a very small group of nations that have not ratified the Basel Convention, which constricts the flow of toxic e-waste to Asia. Currently, all of the Asian countries with burgeoning unregulated reclamation industries are signatories, but containers of e-waste are seldom stopped in transit, and they are hardly ever inspected as they leave America or cross international boundaries. Richard Black, a science writer for the BBC, has documented how easy it still is for America's e-waste contraband to cross into the People's Republic through Hong Kong's container port: "Brokers tell us they carefully tape $100 bills just inside the back of the shipping container so when the customs agents open up these containers, they've got their bribe and it can just pass on through . . . the containers are really full of all sorts of things, including we're told . . . bribes as big as a Mercedes."[35]

Our actions as consumers of electronic goods clearly have ripple effects around the globe. We now live in a complicated world where our consumption of technology must include routine safeguards, and where naive or unconscious consumption of technology is no longer a healthy option. In its 2002 report, *Technically Speaking: Why All Americans Need to Know More about Technology*, the national Committee on Technological Literacy described their mandate in this way: "American adults and children have a poor understanding of the essential characteristics of technology, how it influences society, and how people can and do affect its development . . . Americans are poorly equipped to recognize, let alone ponder or address, the challenges technology poses. And the mismatch is growing. Although our use of technology is increasing . . . there is no sign of an improvement in our ability."[36]

The national Committee on Technological Literacy has begun a reform of America's K-12 educational system to accommodate society's need for a technologically aware populace. Two years before issuing their overview of the problem in *Technically Speaking,* they published their curriculum and content guidelines in *Standards for Technological Literacy: Content for the Study of Technology.*[37] Implementation of new pedagogical content, however, is always slow, and America cannot afford to wait a generation before solving the problem of e-waste and whatever new technological challenges will follow.

During the next few years, the overwhelming problem of waste of all kinds will, I believe, compel American manufacturers to modify industrial practices that feed upon a throwaway ethic. The golden age of obsolescence—the heyday of nylons, tailfins, and transistor radios—will go the way of the buffalo. Whatever comes in its place will depend on the joint effort of informed consumers and responsive manufacturers, who will, I believe, see the benefits of genuinely serving their customers' interests through green design. Very soon, the sheer volume of e-waste will compel America to adopt design strategies that include not just planned obsolescence but planned disassembly and reuse as part of the product life cycle. This is the industrial challenge of the new century. We must welcome it.

Introduction

1. Environmental Health Center, National Safety Council, "Electronic Product Recovery and Recycling Baseline Report," 1999.
2. Maryfran Johnson, "Cleaning IT's Basement," *Computerworld*, February 2, 2004, p. 14.
3. Silicon Valley Toxic Coalition, *Just Say No to E Waste: Background Document on Hazardous Waste from Computers*. Available at http://www.svtc.org/cleancc/pubs/sayno.htm#junk.htm.
4. Eric Most, *Calling All Cell Phones: Collection, Reuse and Recycling Programs in the US* (New York: Inform, 2003), p. 1.
5. Eric Williams, "Environmental Impacts in the Production of Personal Computers," in Ruediger Keuhr and Eric Williams, eds., *Computers and the Environment: Understanding and Managing Their Impacts* (Dordrecht: Kluwer Academic Pubications, 2003), p. 58.
6. Janice Williams Rutherford, *Selling Mrs. Consumer: Christine Frederick and the Rise of Household Efficiency* (Athens: University of Georgia Press, 2003), p. 52.

1. Repetitive Consumption

1. "Production and Consumption," *U.S. Economist and Dry Goods Reporter*, May 6, 1876, p. 7. Cited in William Leach, *Land of Desire: Mer-*

chants, Power and the Rise of a New American Culture (New York: Pantheon, 1993), p. 36.

2. King Camp Gillette, *The People's Corporation* (New York: Boni and Liveright, 1924), p. 237.

3. Archibald Shaw, "Some Problems in Market Distribution," *Quarterly Journal of Economics,* August 1912, pp. 703–765.

4. Edward Filene, *The Way Out* (Garden City, NY: Doubleday, 1925), pp. 196–197.

5. Susan Strasser, *Satisfaction Guaranteed: The Making of the American Mass Market* (New York: Pantheon, 1989), p. 57.

6. Christine Frederick, *Household Engineering* (New York: Business Bourse, 1919), pp. 355–356.

7. Reginald Belfield, "Westinghouse and the Alternating Current, 1935–1937," cited in Jill Jonnes, *Empires of Light: Edison, Tesla, Westinghouse and the Race to Electrify the World* (New York: Random House, 2003), pp. 132–133.

8. Amy K. Glasmeier, *Manufacturing Time: Global Competition in the Watch industry, 1795–2000* (London: Guildord, 2000), p. 121.

9. Russell B. Adams Jr., *King Camp Gillette: The Man and His Wonderful Shaving Device* (Boston: Little, Brown, 1978), p. 7.

10. Tim Dowling, *Inventor of the Popular Culture: King Camp Gillette, 1855–1932* (London: Short Books, 2001), p. 20.

11. Ibid., pp. 33–35.

12. Sinclair Lewis, *Babbitt* (New York: Harcourt Brace, 1922), p. 401.

13. Cited in Janice Williams Rutherford, *Selling Mrs. Consumer: Christine Frederick and the Rise of Household Efficiency* (Athens: University of Georgia Press, 2003), p. 31.

14. There is little consensus about the nature and extent of the changes in women's status, but see Mary Sydney Branch, *Women and Wealth: A Study of the Economic Status of American Women* (Chicago: University of Chicago Press, 1934), p. 119; Frank Presbrey, *The History and Development of Advertising* (New York: Greenwood Press, 1968), p. 567; and S. Mintz and S. Kellogg, *Domestic Revelations: A Social History of American Family Life* (New York: Macmillan, 1988).

15. W. H. Black, *The Family Income* (self-published, 1907), p. 35.

16. Kate Ford, "Celluloid Dreams: The Marketing of Cutex in America, 1916–1935," *Journal of Design History,* 15, no. 3 (2002): 183.

17. Roland Marchand, *Advertising the American Dream* (Berkeley: University of California Press, 1985), p. 54.
18. "Cellucotton," *Fortune Magazine,* November 1937, p. 196.
19. Marchand, *Advertising the American Dream,* pp. 22–24; Susan Strasser, *Waste and Want: A Social History of Trash* (New York: Pantheon, 2001), pp. 160–169.
20. John Gunther, *Taken at the Flood: The Story of Albert D. Lasker* (New York: Harper, 1960), p. 164, cited in Strasser, *Waste and Want,* p. 163.
21. Strasser, *Waste and Want,* p. 174.
22. Edward William Bok, *The Americanization of Edward Bok* (New York: Scribner, 1920), p. 101.
23. John Gould, "Pencils That Didn't Erase, and Chains That Grew," *Christian Science Monitor,* May 16, 2003.
24. Thomas Nixon Carver, *War Thrift: Preliminary Economic Studies of the War,* vol. 10 (1919), p. 23.
25. Henry Ford, *My Life and Work* (New York: Doubleday, 1922), p. 186.
26. Clarence Wilbur Taber, *The Business of the Household,* 2nd ed. (Philadelphia: Lippincott, 1922), p. 438.
27. Malcom Cowley, *Exile's Return: A Literary Odyssey of the 1920s* (New York: Viking, 1951), p. 322.

2. The Annual Model Change

1. Henry Ford, *My Life and Times* (New York: Macmillan, 1922), p. 59.
2. Allan Nevins and Frank Ernest Hill, *Ford: Expansion and Challenge, 1915–1933* (New York: Scribner, 1957), p. 439.
3. Louis L. Sullivan, *Autobiography of an Idea* (New York: Dover, 1924), p. 329. George Basalla, *The Evolution of Technology* (Cambridge: Cambridge University Press, 1988), p. 21.
4. Thorstein Veblen popularized the word in 1899 by using it frequently in *Theory of the Leisure Class.* A characteristic Veblen usage follows: "*Classic* always carries this connotation of wasteful and archaic, whether it is used to denote the dead languages or the obsolete or obsolescent forms of thought and diction in the living language, or to denote other items of scholarly activity or apparatus to which it is applied with less aptness. So the archaic idiom of the English language is spoken of as 'classic' English. Its use is imperative in all speaking and

writing upon serious topics, and a facile use of it lends dignity to even the most commonplace and trivial string of talk. The newest form of English diction is of course never written."

5. Stuart W. Leslie, *Boss Kettering* (New York: Columbia University Press, 1983), p. 138.

6. Peter F. Drucker, *Managing in the Next Society* (New York: St. Martin's, 2002), p. 97.

7. Leslie, *Boss Kettering*, p. 142.

8. David Gartman, *Auto Opium: A Social History of American Automobile Design* (London: Routledge, 1994), p. 75.

9. Philip Ball, *Bright Earth: Art and the Invention of Color* (New York: Farrar, Straus and Giroux, 2002), pp. 225–227.

10. Nevins and Hill, *Expansion and Challenge*, p. 406. Quotation cited in Stella Benson, *The Little World* (New York: Macmillan, 1925), p. 3.

11. Nevins and Hill, *Expansion and Challenge*, p. 151.

12. Preston W. Slosson, *The Great Crusade and After* (New York, 1931), p. 130, cited in Nevins and Hill, *Expansion and Challenge*, p. 399.

13. Cited in Nevins and Hill, *Expansion and Challenge*, p. 400.

14. Gartman, *Auto Opium*, pp. 45–46.

15. The evidence concerning the extent of Pierre S. DuPont's role in developing auto finishes and his intentions in obtaining monopolistic control over them is complicated and was likely obscured during or before by the antitrust investigation against General Motors. K. Ford, "Celluloid Dreams: The Marketing of Cutex in America, 1916–1935," *Journal of Design History*, 15, no. 3 (2002): 175–189.

16. Gary Cross, *Kid's Stuff: Toys and the Changing World of American Childhood* (Cambridge: Harvard University Press, 1977), p. 55.

17. James C. Young, "Ford To Fight It Out with His Old Car," *New York Times*, December, 26, 1926, sec. 8, p. 1.

18. Ibid.

19. M. Lamm, "Harley Earl's California Years, 1893–1927," *Automobile Quarterly*, 20, no. 1 (1982): 34–44.

20. Gartman, *Auto Opium*, p. 95.

21. Alfred P. Sloan Jr., "Alfred Sloan on the Future of the Automobile," *Printer's Ink*, May 3, 1928, pp. 57–58.

22. Jane Fiske Mitarachi, "Harley Earl and His Product: The Styling Section," *Industrial Design*, 2 (October 1955): 52, cited in Gartman, *Auto Opium*, p. 97.

23. *New York Times,* September 29, 1927, cited in Nevins and Hill, *Expansion and Challenge,* p. 438.

24. Nevins and Hill, *Expansion and Challenge,* pp. 455–457.

25. Gartman, *Auto Opium,* p. 77.

26. *Chicago Tribune,* December 1, 1927, p. 9.

27. Gartman, *Auto Opium,* p. 92.

28. Huger Elliott, "The Place of Beauty in the Business World," *Annals of the American Academy of Political and Social Science,* 115 (September 1924): 56.

29. Earnest Elmo Calkins, "Beauty: The New Business Tool," *Atlantic Monthly,* August 1927.

30. Veblen, *Theory of the Leisure Class.*

31. Sunny Y. Auyang, *Engineering—An Endless Frontier* (Cambridge: Harvard University Press, 2004), p. 62.

32. Victor J. Papanek, *Design for Human Scale* (New York: Van Nostrand Reinhold, 1983), pp. 46, 68.

33. Tibor de Skitovsky, *The Joyless Economy: An Inquiry into Human Satisfaction and Consumer Satisfaction* (New York: Oxford University Press, 1976), p. 61: "The drug user's desire for his drug is not qualitatively different from the average person's desire to continue consuming whatever he habitually consumes." See also Helga Ditmar, *The Social Psychology of Material Possessions* (London: Wheatsheaf, 1992). One exception is Jean Kilbourne's *Can't Buy My Love: How Advertising Changes the Way We Think and Feel* [formerly *Deadly Persuasion*] (New York: Touchstone, 1999), p. 233: "Advertising most contributes to the addictive mindset by trivializing human relationships and encouraging us to feel that we are in relationships with our products, especially with those products that are addictive." W. R. Heath, "Advertising That Holds the 'Mauve Decade' Up to Ridicule," *Printer's Ink,* May 10, 1928, p. 77.

34. Emanuel Levy, *Oscar Fever: The History and Politics of the Academy Awards* (New York: Continuum, 2001), p. 1. See also Erica J. Fischer, *The Inauguration of Oscar* (New York: Saur, 1988), and Pierre Norman Sands, *A Historical Study of the Academy of Motion Picture Arts and Sciences, 1927–1947* (New York: Arno, 1973).

35. John Bear, *The #1 New York Times Bestseller: Intriguing Facts about the 484 Books That Have Been #1 New York Times Bestsellers since 1942* (Berkeley: Ten Speed Press, 1992), p. 61.

3. Hard Times

1. Janice Williams Rutherford, *Selling Mrs. Consumer: Christine Frederick and the Rise of Household Efficiency* (Athens: University of Georgia Press, 2003), pp. 25–39, 50–51.

2. J. George Frederick, "Is Progressive Obsolescence the Path toward Increased Consumption?" *Advertising and Selling,* 11, no. 10 (September 5, 1928): 19, 20, 44, 46.

3. Ibid., p. 44.

4. Ibid., p. 49.

5. Joseph A. Schumpeter, *The Theory of Economic Development: An Enquiry into Profits, Capital, Credits, Interest and the Business Cycle,* trans. R. Opie (Cambridge: Harvard University Press, 1934).

6. Eduard März, *Joseph Schumpeter: Scholar, Teacher, Politician* (New Haven: Yale University Press, 1991), p. 5.

7. Schumpeter, *Capitalism, Socialism and Democracy,* pp. 82–83.

8. Paul M. Mazur, *American Prosperity: Its Causes and Consequences* (London: Jonathan Cape, 1928), p. 98.

9. Ibid., p. 99.

10. Frederick, "Is Progressive Obsolescence the Path toward Increased Consumption?" p. 44.

11. W. E. Freeland, "Style and Its Relation to Budgeting," *General Management Series,* 91, pp. 3–19; I. D. Wolf and A. P. Purves, "How the Retailer Merchandises Present-Day Fashion, Style, and Art," *General Management Series,* 97, pp. 2–24; P. Bonner, P. Thomas, J. E. Alcott, and H. E. Nock, "How the Manufacturer Copes with the Fashion, Style and Art Problem," *General Management Series,* 98, pp. 3–23; R. Abercrombie, "The Renaissance of Art in American Business," *General Management Series,* 99, pp. 3–8; E. G. Plowman, "Fashion, Style and Art Spread to Other Lines of Business," *General Management Series,* 106, pp. 3–32.

12. Christine M. Frederick, *Selling Mrs. Consumer* (New York: Business Bourse, 1929), p. 245.

13. Gwendolyn Wright, cited in Rutherford, *Selling Mrs. Consumer,* p. 195.

14. Ibid., p. 4.

15. Frederick, *Selling Mrs. Consumer,* p. 246.

16. Justus George Frederick, *A Philosophy of Production* (New York: Business Bourse, 1930), p. 230.

17. "Peed Tells Why New Cars Show Radical Change: 'Progressive Obso-

lescence' Is the Key," *Advertising Age,* January 13, 1934. This is the only use of Frederick's phrase until the 1950s that I have been able to find.

18. Sigfried Giedion, *Mechanization Takes Command: A Contribution to Anonymous History* (New York: Oxford University Press, 1948), p. 610.

19. Roy Sheldon and Egmont Arens, *Consumer Engineering: A New Technique for Prosperity* (New York: Harper and Brothers, 1932), pp. 13–14.

20. Frederick, *Selling Mrs. Consumer,* p. 246.

21. Sheldon and Arens, *Consumer Engineering,* p. 56.

22. Ibid., p. 54.

23. Ibid., p. 7.

24. Kerry Seagrave, *Vending Machines: An American Social History* (Jefferson, NC: McFarland, 2002), p. 106. Silver Sam, "Coin Operated Amusement Machines," *Billboard,* April 23, 1932, p. 67.

25. "Obsolete Men," *Fortune,* 6 (December 1932): 25, 26, 91, 92, 94. This idea would eventually receive its highest expression in Günther Anders's 1956 philosophical treatise *The Obsolescence of Man.* Writing in German twenty years after immigrating to the United States, Anders predicted the complete replacement of humans by machines and the end of human history. He saw the Holocaust as the first attempt at systematic extermination of a whole people by industrial means. Anders was probably not influenced by MacLeish. It is more likely that he read Norman Cousins's famous essay "Modern Man Is Obsolete," which followed the attacks on Hiroshima and Nagasaki in August 1945 (see Chapter 5).

26. Archibald MacLeish, *New and Collected Poems, 1917–1976* (Boston: Houghton Mifflin, 1976), p. 295, "Invocation To The Social Muse."

27. MacLeish, "Obsolete Men," p. 92.

28. Ibid., p. 94.

29. Ibid., p. 92.

30. Rautenstrauch was a follower of Henry L. Gantt, Frederick Taylor's most successful disciple. In 1917 he joined Gantt in one of technocracy's prototypes, a group called The New Machine, which appealed to President Wilson to put industry into the hands of those who understood it best. For more about Rautenstrauch, see Edwin T. Layton Jr., *The Revolt of the Engineers: Social Responsibility and the American Engineering Profession* (Baltimore: Johns Hopkins University Press, 1971), pp. 147, 227, 228.

31. William E. Atkin, *Technocracy and the American Dream: The Technocrat Movement, 1900–1941* (Berkeley: University of California Press, 1977), pp. 59–60.

32. Henry Elsner Jr., *The Technocrats: Prophets of Automation* (Syracuse: Syracuse University Press, 1967), p. 7. The term "technocracy" comes from Dr. Virgil Jordan, president of the National Industrial Conference Board, cited in ibid., p. 8.

33. Atkin, *Technocracy and the American Dream*, pp. 64–65.

34. Elsner, *Technocrats: Prophets of Automation*, pp. 8–11.

35. Ibid., p. 13.

36. For London's dates and other information concerning his Masonic affiliation I am indebted to Tom Savini, director, Chancellor Robert R. Livingston Lodge, New York, New York, personal correspondence [email message], March 11, 2005.

37. Elsner, *Technocrats: Prophets of Automation*, p. 3. Copies of London's work are now extremely rare, and because they are not well known, they are still not very valuable. The work is generally available, however, since London had sufficient ambition for his pamphlet to copyright it by registering it with the Library of Congress on September 15, twenty-five days after Howard Scott achieved national recognition through his first interview in the *New York Times*. London's acquaintance with the technocrats can be assumed, since he would not have enough time to write his pamphlet, print it, mail it to the LC, and have it processed for copyright simply after reading the *New York Times* interview. I am considerably indebted both for her generosity and friendship to the Library of Congress's gifted reference librarian, Emily Howie, for patiently digging out and photocopying one of the few survival original copies of London's first pamphlet. In total, Bernard London wrote three essays: *Ending the Depression Through Planned Obsolescence* (New York: self-published, 1932), 2 pp.; *The New Prosperity through Planned Obsolescence: Permanent Employment, Wise Taxation and Equitable Distribution of Wealth* (New York: self-published, 1934), 67 pp.; and *Rebuilding a Prosperous Nation through Planned Obsolescence* (New York: self-published, 1935), 40 pp. All are available at the Library of Congress.

38. London, *Ending the Depression through Planned Obsolescence*.

39. Ibid., pp. 6–7.

40. Ibid., p. 12.

41. Ibid., p. 13.

42. Aldous Huxley, *Brave New World* (Harmondsworth: Penguin, 1959), p. 49.

43. Ibid., p. 35.

44. Susan Strasser, *Waste and Want: A Social History of Trash* (New York: Henry Holt, 1999), pp. 274–278.

45. Sigfried Giedion, *Mechanization Takes Command: A Contribution to Anonymous History* (New York: Oxford University Press, 1948), pp. 181.

46. J. A. Hobson, *Work and Wealth: A Human Valuation* (New York: Kelley, 1968), p. 90.

47. Stuart Chase, *The Tragedy of Waste* (New York: Macmillan, 1925), pp. 70, 74.

48. Leon Kelley, "Outmoded Durability: If Merchandise Does Not Wear Out Faster, Factories Will Be Idle, People Unemployed," *Printers' Ink*, January 9, 1936.

49. Lewis Mumford, *Technics and Civilization* (New York: Harcourt Brace, 1934), p. 394.

50. George Wise, *Willis R. Whitney, General Electric and the Origins of U.S. Industrial Research* (New York: Columbia University Press, 1985), p. 117.

51. Ibid., p. 41.

52. Letter from L. C. Porter to M. I. Sloan, November 1, 1932. U.S. v G.E. Civil Action No. 1364 Ex. 1860-G. Letter quoted in George W. Stocking and Myron W. Watkins, *Cartels in Action: Case Studies in International Business Diplomacy* (New York: Twentieth Century Fund, 1946), p. 354. Original reference found in Vance Packard, *The Waste Makers* (New York: David McKay, 1960), p. 59.

53. Ibid.

4. Radio, Radio

1. Pupin's autobiography was a bestseller in its day. For more about his spirituality and "scientific idealism," see Edwin T. Layton, *The Revolt of the Engineers: Social Responsibility and the American Engineering Profession* (Baltimore: Johns Hopkins University Press, 1971), pp. 39, 49,

92, and esp. 216; see also A. Michael McMahon, *The Making of a Profession: A Century of Electrical Engineering in America* (New York: IEEE Press, 1984), pp. 44, 47–48, 51, 123, 134–135, and 138.

2. Layton, *Revolt of the Engineers,* p. 212.

3. Cited in Tom Lewis, *Empire of the Air: The Men Who Made Radio* (New York: Harper Collins, 1991), p. 178.

4. Lawrence Lessing, *Man of High Fidelity: Edwin Howard Armstrong, a Biography* (Philadelphia: Lippencott, 1956), p. 212.

5. Eugene Lyons, *David Sarnoff, a Biography* (New York: Harper & Row, 1966), p. 213.

6. Cited in Lewis, *Empire of the Air,* p. 278.

7. George Wise, *Willis R. Whitney, General Electric, and the Origins of U.S. Industrial Research* (New York: Columbia University Press, 1985), p. 219.

8. Lewis, *Empire of the Air,* p. 261.

9. Ibid., pp. 259–260.

10. This was the same antitrust case that had revealed GE was experimenting with shortened death dates (see Chapter 3).

11. Lyons, in *David Sarnoff, a Biography,* p. 219, claims that Fly had given Sarnoff tacit approval for the sale of these sets. RCA planned to build 25,000 of them, but—according to Steve McVoy of the Early Television Foundation—only about 3,400 of them were manufactured before production ended in 1942. What is probably true about the incident with Fly is that Sarnoff suggested the idea to Fly without specifying any numbers or dates, and Fly agreed in principle but became angry later when he realized with what liberty Sarnoff had interpreted his assent.

12. Lewis, *Empire of the Air,* p. 300.

13. This investigation was organized by Sen. Burton K. Wheeler under the auspices of the Committee on Interstate Commerce. Both Fly and Sarnoff testified; see Lyons, *David Sarnoff, a Biography,* p. 219.

14. Cited in Lessing, *Man of High Fidelity,* p. 244.

15. Cited in Lewis, *Empire of the Air,* p. 277.

16. Joseph A. Schumpeter, *Capitalism, Socialism and Democracy* (New York: Harper and Brothers, 1942), pp. 131–134.

17. Lyons, *David Sarnoff, a Biography,* p. 220.

18. "Development of Television, Hearings on S. Res. 251, A Resolution Re-

questing the Committee on Interstate Commerce to Investigate the Actions of the Federal Communications Commission in Connection with the Development of Television." Hearings before the Senate Committee on Interstate Commerce (76th Congress, 3rd Session), April 10 and 11, 1940, p. 49. Cited in Lessing, *Man of High Fidelity,* p. 244.

19. Hearings, p. 32.
20. I am indebted to Steve McVoy at the Early Television Foundation for this technical information. He credits the actual production figures of TRKs to Jeff Lendaro.
21. Lewis, *Empire of the Air,* p. 288.
22. Ibid., p. 285.
23. "Progress of FM Radio: Hearings before the Senate Committee on Interstate and Foreign Commerce on Certain Changes Involving Development of FM Radio and RCA Patent Policies," March 30, April 23, May 12, 13, 21, 1948 (378 pp.).
24. Lyons, *David Sarnoff, a Biography,* p. 112.
25. Michael Brian Schiffer, *The Portable Radio in American Life* (Tucson: University of Arizona Press, 1991), pp. 161–162.
26. Ibid.
27. Paul Eisler, *My Life with the Printed Circuit* (Bethlehem, PA: Lehigh University Press, 1989), p. 170.
28. Alexander Feklisov and Sergei Kostine, *The Man behind the Rosenbergs* (New York: Enigman Books, 2001), p. 121.
29. Schiffer, *Portable Radio in America,* pp. 162–169. Professor Schiffer is the proud possessor of a working Belmont.
30. Otto J. Scott, *The Creative Ordeal: The Story of Raytheon* (New York: Atheneum, 1974), p. 117.
31. Michael Riordan and Lillian Hoddeson, *Crystal Fire: The Birth of the Information Age* (New York: Norton, 1997), p. 1.
32. Ibid., p. 6.
33. Micheal Schiffer points out that tubes are still used in some specialized applications like CRTs and the highest power transmitters. As is so often true, obsolescence is not an absolute value. Obsolete devices often survive for specialized uses.
34. Described in John Dielbold, *Automation: The Advent of the Automatic Factory* (New York: Van Nostrand, 1952), p. 39.

5. The War and Postwar Progress

1. Martin Booth, *Opium: A History* (New York: Simon & Schuster, 1996), pp. 161, 163.
2. Ibid., p. 287.
3. "Tightening of Economic Noose Puts Next Move Up to Japan," *Newsweek*, August, 11, 1941, p. 16.
4. *Pearl Harbor Attack: Hearings before the Joint Committee on the Investigation of the Pearl Harbor Attack* (Washington, DC: U.S. Government Printing Office, 1946), p. 15.
5. Bruce Barton, *The Man Nobody Knows* (London: Constable, 1925), p. 103.
6. Elmer K. Bolton, "Annual Report," January 2, 1934, Hagley Museum Library Collection, Wilmington, DE, 1784, Box 16. Cellulose rayon producers had developed and patented means for making rayon more marketable as a substitute for other textiles by dulling its lustre with the addition of fine particles of titanium dioxide to the polymer. DuPont did not want to go to the expense of licensing this process from its competitors.
7. Wallace Carothers to Hamilton Bradshaw, November 9, 1927, Hagley Museum and Library Collection, Wilmington, DE, 1896, cited in Matthew E. Hermes, *Enough for One Lifetime: Wallace Carothers, Inventor of Nylon* (Washington, DC: American Chemical Society and Chemical Heritage Foundation, 1996), p. 56.
8. Wallace Carothers to John R. Johnson, January 9, 1934, cited in Hermes, *Enough for One Lifetime*, p. 184.
9. Cited in Hermes, *Enough for One Lifetime*, p. 187.
10. Madeline Edmonson and David Rounds, *The Soaps: Daytime Serials of Radio and TV* (New York: Stein and Day, 1973), pp. 38–39.
11. Cited in Jeffrey Meikle, *American Plastic: A Cultural History* (Rutgers: Rutgers University Press, 1995), p. 137.
12. "$10,000,000 Plant to Make Synthetic Yarn: Major Blow to Japan's Silk Trade Seen," *New York Times*, October 21, 1938, p. 1.
13. Cited in Susannah Handley, *Nylon: The Story of a Fashion Revolution* (Baltimore: Johns Hopkins University Press, 1999), p. 35, and therein attributed to Stephen Fenichell, *Plastic: The Making of a Synthetic Century* (New York: Harper Collins, 1996), p. 139.
14. Handley, *Nylon: Story of a Fashion Revolution*, p. 37.

15. Nicols Fox, author of *Against the Machine: The Hidden Luddite Tradition in Literature, Art and Individual Lives* (Washington, DC: Island Press, 2002), has this memory of her father, Julian Powers Fox, a Virginia Polytechnic Institute engineering graduate, class of 1933: "I can remember quite distinctly my father, who was an engineer and worked for DuPont in the nylon division in those early days, telling me how the early nylon stockings were too durable and had to be made more fragile so that women would have to buy them more frequently. The 'too good' stockings were bad for business. As someone who resented the money they cost and suffered the frustrations of often getting a run in one from simply pulling them on, this made quite an impression—probably turned me into a Democrat at that precise moment." I am grateful to Nicols Fox for suggesting nylon as a topic for this section.

16. Theodore Levitt, *The Marketing Imagination* (New York: Macmillan, 1983), p. 190.

17. Milton N. Grass, *History of Hosiery from the Piloi of Ancient Greece to the Nylons of Modern America* (New York: Fairchild, 1955), p. 264.

18. Alexander Felisov and Sergei Kostine, *The Man behind the Rosenbergs* (New York: Enigma Books, 2001), p. 55.

19. William L. Langer and S. Everett Gleason, *The Undeclared War, 1940–1941* (New York: Harper, 1953), p. 708.

20. John O'Riley, "Japan: Its Industries Live on Borrowed Time," *Wall Street Journal,* December 2, 1941, p. 3.

21. *New York Times,* "Trade Winds," December 7, 1941, sec. 4, p. 3.

22. Michael Brian Schiffer, "The Explanation of Long-Term Technological Change," in Michael Brian Schiffer, ed., *Anthropological Perspectives on Technology* (Albuquerque: University of New Mexico Press, 2001), p. 217.

23. Kenneth T. Jackson, *Crabgrass Frontier: The Suburbanization of the United States* (New York: Oxford University Press, 1985), p. 197.

24. Rosalyn Baxandall and Elizabeth Ewen, *Picture Windows: How the Suburbs Happened* (New York: Basic Books, 2000), p. 87.

25. Ibid.

26. E. W. Burgess, "The Growth of the City," in Robert E. Park, Ernest W. Burgess, and Roderick D. McKenzie, *The City* (Chicago: University of Chicago Press, 1925), p. 159.

27. Tom Bernard, "New Homes for Sixty Dollars a Month," *American*

Magazine, April 1948, p. 105. Cited in Baxandall and Ewen, *Picture Windows,* p. 125.

28. "Two Publishers Get Prison Sentences," *New York Times,* June 14, 1939, p. 8.

29. S. Duncan Black and Alonzo G. Decker produced the world's first hand-held power tool, a drill, in 1914. Belt sanders were available in America by 1924. As far as I know there is not yet a complete history of power tools and their substantial impact on American society, although this would make a wonderful study.

30. John Thomas Leill, "Levittown: A Study in Community Planning and Development," Ph.D. diss., Yale University, 1952, p. 61.

31. Jackson, *Crabgrass Frontier,* pp. 237–238.

32. Michael Dolan, *The American Porch: An Informal History of an Informal Place* (New York: Lyons, 2002).

33. Sharon Zukin, *The Cultures of Cities* (Cambridge: Blackwell, 1995).

34. Tom Martinson, *American Dreamscape: The Pursuit of Happiness in Postwar Suburbia* (New York: Carroll and Graf, 2000), p. 137.

35. Barbara M. Kelly, *Expanding the American Dream: Building and Rebuilding Levittown* (Albany: State University of New York Press, 1993), p. 69.

36. "Levitt's Progress," p. 168.

37. Mark Kurlansky, *Salt: A World History* (New York: Vintage, 2002), p. 307.

38. John Dean, *Home Ownership: Is It Sound?* (New York: Harper & Row, 1945), p. 25. Cited in Baxandall and Ewen, *Picture Windows,* p. 135.

39. See Konrad Zuse, *Der Computer, mein Lebenswerk* (Munich, 1970).

40. Alice R. Burks and Arthur W. Burks, *The First Electronic Computer: The Atansoff Story* (Ann Arbor: University of Michigan Press, 1988), p. 124.

41. Larry Owens, "Vannevar Bush and the Differential Analyzer: The Texts and Context of an Early Computer," *Technology and Culture,* 27, no. 1 (January 1986): 93–94.

42. Ibid., p. 64.

43. Scott McCartney, *ENIAC: The Triumphs and Tragedies of the World's First Computer* (New York: Walker, 1999), p. 34.

44. Ibid.

45. Ibid., p. 36.

46. John Gustafson, "Reconstruction of the Atanasoff-Berry Computer," in Raul Rojas and Ulf Hashagen, eds., *The First Computers: History and Architectures* (Cambridge: MIT Press, 2000), p. 101.

47. Cited in Clark R. Mollenhoff, *Atanasoff: Forgotten Father of the Computer* (Ames: Iowa State University Press, 1988), pp. 3, 70, 59.

48. McCartney, *ENIAC*, pp. 51, 56.

49. Warren Weaver's project diaries for March 17, 1950, cited in Owens, "Vannevar Bush and the Differential Analyzer," p. 64.

50. Cited in Mollenhoff, *Atanasoff*, p. 8.

51. Paul Boyer, *By The Bomb's Early Light: American Thought and Culture at the Dawn of the Atomic Age* (New York: Pantheon, 1985), p. 40.

52. Norman Cousins, "Modern Man Is Obsolete," *Saturday Review*, August 18, 1945.

53. J. Samuel Walker, *Prompt and Utter Destruction: Truman and the Use of Atomic Weapons against Japan* (Chapel Hill: University of North Carolina Press, 1997), p. 98.

54. Cited in Boyer, *By The Bomb's Early Light*, p. 10.

55. Cousins, "Modern Man Is Obsolete."

56. Wendell Lewis Willkie, *One World* (New York: Simon & Schuster, 1943).

57. Cousins, "Modern Man Is Obsolete."

58. Boyer, *By The Bomb's Early Light*, p. 40.

59. Norman Cousins, "The Standardization of Catastrophe," *Saturday Review*, August 10, 1946, pp. 16–18.

60. Lewis L. Strauss, *Men and Decisions* (London: Macmillan, 1963), p. 209.

61. Cousins, "Standardization of Catastrophe," pp. 16–18.

62. Norman Cousins, "Bikini's Real Story," *Saturday Review*, December 11, 1948, p. 15.

63. Boyer, *By The Bomb's Early Light*, p. 44.

6. The Fifties and Sixties

1. Quoted in Glenn Adamson, *Industrial Strength Design: How Brooks Stevens Changed Your World* (Cambridge: MIT Press, 2003), p. 145.

2. Ibid., pp. 46, 79.

3. John Heskett, "The Desire for the New: The Context of Brooks Steven's Career," in Adamson, *Industrial Strength Design*, pp. 2–3.

4. Ron Grossman, *"The Idea Man," Chicago Tribune,* June 3, 1991, cited in Heskett, *"The Desire for the New,"* p. 8.
5. Ibid.
6. Leon Kelley, "Outmoded Durability: If Merchandise Does Not Wear Out Faster, Factories Will Be Idle, People Unemployed," *Printers' Ink,* January 9, 1936.
7. Cited in Heskett, *"The Desire for the New,"* p. 4.
8. Brooks Stevens Associates, untitled brochure, 1953. Brooks Stevens Archive cited in Heskett, *"The Desire for the New,"* p. 4.
9. Kristina Wilson, "Brooks Stevens, The Man in Your Life: Shaping the Domestic Sphere, 1935–1950," in *Industrial Strength Design,* p. 9.
10. Many articles were self-published and are now inaccessible, but a selection of them has been conveniently reissued in the appendixes of Adamson's thorough and fascinating *Industrial Strength Design.*
11. Karl Prentiss, "Brooks Stevens: He Has Designs on Your Dough," *True: The Man's Magazine,* April 1958, cited in Heskett, *"The Desire for the New,"* p. 4.
12. Available online from the United States Department of Transportation website at http://www.fhwa.dot.gov/infrastructure/hearst.htm. Downloaded April 10, 2005.
13. David Gartmann, *Auto Opium: A Social History of American Automobile Design* (London: Routledge, 1994), p. 144.
14. Ibid., p. 150.
15. For more information about the Edsel's creation and failure, see Thomas E. Bonsall, *Disaster in Dearborn: The Story of the Edsel* (Stanford: Stanford University Press, 2002), esp. chap. 12, "Why the Edsel Failed."
16. Herbert Marshall McLuhan, *Understanding Media: The Extensions of Man* (New York: McGraw-Hill, 1964), p. 338.
17. George Nelson, "Obsolescence," *Industrial Design,* 3, no. 6 (December 1956): 81–82, 86–89; or see George Nelson, *Problems of Design* (New York: Whitney Publications, 1957), p. 42.
18. For Sottsass's touching personal memoir of George Nelson see his foreword to Stanley Abercrombie, *George Nelson: The Design of Modern Design* (Cambridge: MIT Press, 1995), pp. vii–xii.
19. Victor J. Papanek, *The Green Imperative* (New York: Thames and Hudson, 1995), p. 55.
20. Nelson, "Obsolescence," p. 88.

21. Colin Campbell, *The Romantic Ethic and the Spirit of Modern Consumerism* (Oxford: Basil Blackwell, 1987), p. 46.

22. Packard, *The Hidden Persuaders*, p. 21.

23. Daniel Horowitz, *Vance Packard and American Social Criticism* (Chapel Hill: University of North Carolina Press, 1994), p. 21.

24. Vance Packard, *The Waste Makers* (New York: David McKay, 1960), pp. 54–55.

25. Barbara Ehrenreich, *Fear of Falling: The Inner Life of the Middle Class* (New York: Pantheon, 1989), p. 18.

26. Packard, *The Waste Makers*, p. 8.

27. Horowitz, *Vance Packard and American Social Criticism*, p. 123.

28. "Is 'The Waste Makers' a Hoax?" *Printers' Ink,* September 30, 1960, pp. 20–22, 24–25, 28–29.

29. "Packard Hoodwinks Most Reviewers," *Printers' Ink,* October 21, 1960, pp. 58–60.

30. "Has Packard Flipped?" *Printers' Ink,* March 10, 1961, p. 67.

31. Horowitz, *Vance Packard and American Social Criticism*, p. 157.

32. Cited in Todd Gitlin, *The Sixties: Years of Hope, Days of Rage* (New York: Bantam, 1987), p. 73.

33. The complete text of the SDS's "Port Huron Statement" is available at *The Sixties Project* webpage: http://lists.village.virginia.edu/sixties/HTML_docs/Resources/Primary/Manifestos/SDS_Port_Huron.html.

34. Packard writes about the planned obsolescence debate in *Design News* in *The Waste Makers*, pp. 63–67.

35. E. S. Stafford, "Product Death Dates—A Desirable Concept?" *Design News,* 13, no. 24 (November 24, 1958): 3.

36. I am indebted to Greg Key, the Commerce Department's archivist, for this information.

37. Stafford, "Product Death Dates—A Desirable Concept?" p. 3.

38. Ibid.

39. Harold L. Chambers, *Design News,* 14, no. 2 (January 19, 1959): 2. All letters quoted are from the "Sound Board" (letters) section of two issues of *Design News:* January 5 and January 19, 1959, pp. 2–3, 149.

40. Ernest R. Cunningham, "Daggers to Death-Dates," *Design News,* 14, no. 2 (January 19, 1959): 149.

41. Jack Waldheim, "Lollipops and Faucets," *Design News,* 4, no. 3 (February 2, 1959): 3.

42. "Planned Obsolescence—Is It Fair? Yes! Says Brooks Stevens; No! Says Walter Dorwin Teague," *Rotarian,* February 1960.
43. Thomas Frank, *The Conquest of Cool: Business Culture, Counterculture, and the Rise of Hip Consumerism* (Chicago: University of Chicago Press, 1997), p. 67.
44. Cited in David Kelly, *Getting the Bugs Out: The Rise, Fall, and Comeback of Volkswagen in America* (New York: Wiley, 2002), p. 88.
45. Jerry Della Femina, *From Those Wonderful Folks Who Gave You Pearl Harbor: Front Line Dispatches from the Advertising War* (New York: Simon & Schuster, 1970), p. 27.
46. Cited in Frank, *The Conquest of Cool,* p. 197.
47. Larry Dobrow, *When Advertising Tried Harder: The Sixties, the Golden Age of American Advertising* (New York: Friendly Press, 1984), p. 9.
48. *Life,* February 10, 1961.
49. Cited in Kelly, *Getting the Bugs Out,* pp. 90, 95.
50. Ad cited in Frank, *The Conquest of Cool,* p. 64.
51. Ibid., p. 31; see also p. 68.
52. Cited in ibid., p. 271n.3.
53. Cited in ibid., p. 209.
54. Cited in ibid., p. 197.
55. Gerome Ragni and James Rado, *Hair: The American Tribal Love-Rock Musical* (New York: Simon & Shuster, 1969), p. 69.
56. Cited in Frank, *The Conquest of Cool,* p. 143.
57. John Kenneth Galbraith, *The Affluent Society* (Boston: Houghton Mifflin, 1958), pp. 210–211.
58. This essay, originally published by the *Harvard Business Review* in 1960, has since been reissued as chap. 8 in Levitt's expanded edition of *The Marketing Imagination* (New York: Macmillan, 1983), pp. 141–172.
59. Ibid., p. 155.
60. Reissued in Levitt, *The Marketing Imagination,* pp. 173–199. Sak Onkvist and John J. Shaw, *Product Life Cycles and Product Management* (Westport, CA: Quorum Books, 1989), p. 121.
61. I. Boustead, *The Milk Bottle* (London: Open University Press, 1972). This study supplied the impetus behind the life cycle examinations of beverage containers in many nations over the next few years.
62. McLuhan, *Understanding Media,* p. 24.

63. Herbert Marshall McLuhan, *The Gutenberg Galaxy* (Toronto: University of Toronto Press, 1962), p. 24.
64. Ibid., pp. 185, 42.
65. McLuhan, *Understanding Media*, p. 230.
66. Ibid., p. 222.
67. Ibid., p. 223.

7. Chips

1. James Martin, *The Computerized Society* (Englewood Cliffs, NJ: Prentice Hall, 1970), p. 539.
2. James Martin, *The Wired Society* (Englewood Cliffs, NJ: Prentice Hall, 1978), p. 15.
3. Paul E. Ceruzzi, *A History of Modern Computing* (Cambridge: MIT Press, 2003), pp. 64, 145, 148. Alan Stone, *Wrong Numbers: The Break Up of AT&T* (New York: Basic, 1989), pp. 200, 145.
4. Ibid.
5. Robert Slater, *Portraits in Silicon* (Cambridge: MIT Press, 1987), p. 211. Joel Shurkin, *Engines of the Mind: The Evolution of the Computer from Mainframes to Microprocessors* (New York: Norton, 1996), p. 305.
6. Ceruzzi, *History of Modern Computing*, p. 135. Slater, *Portraits in Silicon*, p. 209.
7. Shurkin, *Engines of the Mind*, p. 136. Slater, *Portraits in Silicon*, p. 212. Ceruzzi, *History of Modern Computing*, p. 149.
8. Slater, *Portraits in Silicon*, p. 167.
9. Ceruzzi, *History of Modern Computing*, pp. 183–184.
10. Slater, *Portraits in Silicon*, p. 156.
11. Ibid., p. 171. Ceruzzi, *History of Modern Computing*, p. 190.
12. Slater, *Portraits in Silicon*, p. 159. Ceruzzi, *History of Modern Computing*, pp. 187–188.
13. Ceruzzi, *History of Modern Computing*, pp. 185, 187.
14. Ibid., pp. 187–188.
15. Gordon Moore, "Cramming More Components on Integrated Circuits," *Electronics*, April 19, 1965, p. 119.
16. Ceruzzi, *History of Modern Computing*, p. 189. In August 1975, two months after Apollo 18 and Soyuz 19 returned to earth, the Soviets

unveiled their own scientific pocket calculator, the Elektronika B3-18. The Elektronika B3-18 was an especially symbolic milestone, since the Soviet space program that launched Sputnik had begun entirely on Sergei Korolev's slide rule. In addition to designing the Semyorka ICBM that launched Sputnik 1 and 2, he designed Yuri Gargarin's Vostok launcher. During incarceration in a Soviet prison, Korolev had became such a virtuoso with his German-made Nestler slide rule that Soviet space engineers referred to it as "the magician's wand."

17. Raymond Kurzweil, *The Age of Spiritual Machines: When Computers Exceed Human Intelligence* (New York: Viking, 1999), pp. 21, 25. James B. Murray Jr., *Wireless Nation: The Frenzied Launch of the Cellular Revolution in America* (Cambridge: Perseus Publishing, 2001), p. 319.

18. Florian Cajori, *A History of the Logarithmic Slide Rule* (1909; New York: Astragal Press, 1991). See especially Cajori's "Addenda" in which he attributes the slide rule's invention to William Oughtred based on evidence in Oughtred's paper "The Circles of Proportion and the Horizontal Instrument" (1632).

19. Slater, *Portraits in Silicon*, p. 55.

20. Dwayne A. Day, John M. Logsdon, and Brian Lattell, eds., *Eye in the Sky: The Story of the Corona Spy Satellites* (Washington: Smithsonian Institute Press, 1998), p. 201. Henry Petroski, *To Engineer Is Human: The Role of Failure in Successful Design* (New York: St. Martin's, 1985), p. 191.

21. Slater, *Portraits in Silicon*, pp. 178–179.

22. Ibid., pp. 178–181.

23. Frederick Seitz and Norman G. Einspruch, *Electronic Genie: The Tangled History of Silicon* (Urbana: University of Illinois Press, 1998), pp. 229–230.

24. The simultaneous development of digital timepieces is described in Amy K. Glasmeier, *Manufacturing Time: Global Competition and the Watch Industry, 1795–2000* (New York: Guildford Press, 2000), pp. 203–241.

25. Ceruzzi, *History of Modern Computing*, p. 213.

26. Petroski, *To Engineer Is Human*, p. 192.

27. Victor J. Papanek, *The Green Imperative* (New York: Thames and Hudson, 1995), p. 175.

28. Tracy Kidder, *The Soul of a New Machine* (Boston: Little, Brown, 1981), p. 59. Petroski, *To Engineer Is Human*, p. 192.

29. Ceruzzi, *History of Modern Computing*, p. 215. Martin, *The Wired Society*, p. 15.

30. Andrea Butter and David Pogue, *Piloting Palm: The Inside Story of Palm, Handspring, and the Birth of the Billion Dollar Handheld Industry* (New York: John Wiley, 2002), p. 305.

31. Victor J. Papanek, *Design for the Real World: Human Ecology and Social Change* (New York: Pantheon, 1972), p. 322. Slater, *Portraits in Silicon*, p. 287.

32. Martin Campbell-Kelly and William Aspray, *Computer: A History of the Information Machine* (New York: Basic Books, 1996), pp. 250, 251.

33. Slater, *Portraits in Silicon*, pp. 285–286.

34. Campbell-Kelly and Aspray, *Computer: A History of the Information Machine*, p. 252. Roy A. Allan, *A History of the Personal Computer: The People and the Technology* (London: Allan Publishing, 2001), p. 17. Shurkin, *Engines of the Mind*, p. 311.

35. J. C. Herz, *Joystick Nation: How Videogames Ate Our Quarters, Won Our Hearts, and Rewired Our Minds* (New York: Little, Brown, 1997), p. 6.

36. Campbell-Kelly and Aspray, *Computer: A History of the Information Machine*, p. 252.

37. Ibid.

38. Ibid., pp. 264–265.

39. Ceruzzi, *History of Modern Computing*, pp. 260, 268–269. Adele Goldberg, *A History of Personal Workstations* (New York: ACM Press, 1988), pp. 195–196. Campbell-Kelly and Aspray, *Computer: A History of the Information Machine*, pp. 266–267.

40. Ceruzzi, *History of Modern Computing*, p. 261.

41. Ibid., p. 211. Stephen Levy, *Insanely Great* (New York: Penguin, 1994), p. 61.

42. The Star workstation disaster has been described in fascinating (if uncharitable) detail in Douglas Smith and Robert Alexander, *Fumbling the Future: How Xerox Invented Then Ignored the First Personal Computer* (New York: William Morrow, 1988).

43. Campbell-Kelly and Aspray, *Computer: A History of the Information Machine*, p. 270. Ceruzzi, *History of Modern Computing*, p. 274.

44. Campbell-Kelly and Aspray, *Computer: A History of he Information Machine*, pp. 270–272.

45. Ibid., pp. 272, 273.

46. Campbell-Kelly and Aspray, *Computer: A History of the Information*

Machine: "Long obsolete, those original Macintoshes are now much sought after by collectors."

47. Ibid., p. 276.
48. Jim Carlton, *Apple: The Inside Story of Intrigue, Egomania, and Business Blunders* (New York: Random House, 1997), p. 132.
49. Campbell-Kelly and Aspray, *Computer: A History of the Information Machine,* p. 278.
50. Ibid., pp. 279–281.
51. Steven Johnson, *Interface Culture: How New Technology Transforms the Way We Create and Communicate* (New York: HarperCollins, 1997), p. 50.
52. Alan Kay, "User Interface: A Personal View," in Brenda Laurel, ed., *The Art of Human-Computer Interface Design* (New York: Addison Wesley, 1990), p. 189. Johnson, *Interface Culture,* p. 50.
53. Ibid. Herbert Marshall McLuhan, *Understanding Media: The Extensions of Man* (New York: McGraw Hill, 1964), p. 24.
54. Herz, *Joystick Nation,* pp. 5–8.
55. Scott Cohen, *Zap! The Rise and Fall of Atari* (New York: McGraw-Hill, 1984), p. 105.
56. Ibid., p. 31.
57. Ibid., pp. 29–30.
58. Ibid., pp. 32–33, 51–52.
59. Ibid.
60. Van Burnham, *Supercade: A Visual History of the Video Game Age, 1971–1984* (Cambridge: MIT Press, 2001), p. 170. Allan, *History of the Personal Computer,* pp. 7, 10. Cohen, *Zap!* p. 70.
61. Ibid. David Sheff, *Game Over: How Nintendo Zapped an American Industry, Captured Your Dollars, and Enslaved Your Children* (New York: Random House, 1993), pp. 8, 128, 129, 296–348.
62. Gary Cross, *Kid's Stuff: Toys and the Changing World of American Childhood* (Cambridge: Harvard University Press, 1997), p. 221. Sheff, *Game Over,* p. 343.
63. Kerry Segrave, *Vending Machines: An American Social History* (Jefferson, NC: McFarland, 2002), p. 172.
64. *The Economist,* 354 (March 2000): 72.
65. Ibid. Cross, *Kid's Stuff,* p. 221.
66. Sheff, *Game Over,* p. 212.

67. Ibid. Larry Downes and Chunka Mui, *Unleashing the Killer App: Digital Strategies for Market Dominance* (Boston: Harvard Business School Press, 1998), p. 158.

8. Weaponizing Planned Obsolescence

1. Victor J. Papanek, *Design for the Real World* (New York: Bantam, 1971), p. 97.
2. Told by Jane Greenbaum Eskind to Harris A. Gilbert, relayed to me in personal correspondence (letter), February 16, 2005; more details supplied by Harris A. Gilbert, personal correspondence (email message), February 22, 2005.
3. *Aviation Week and Space Technology,* August 29, 1977, p. 19; June 7, 1976, p. 13.
4. Harris A. Gilbert, personal correspondence (letter), February 22, 2005. Details of Gus's painful childhood from Dr. Eric M. Chazen, personal correspondence (email message), January 11, 2005. Laurie Weisman, Gus's cousin and executrix (telephone interview), March 30, 2005.
5. Maurice Eisenstein, personal correspondence (letter), February 22, 2005; Arnold Kramish, personal correspondence (email message), January 16, 2005.
6. Maurice Eisenstein, personal correspondence (personal letter) February 22, 2005; Dr. Eric M. Chazen, personal correspondence (email message), January 12, 2004. Derek Leebaert, *The Fifty-Year Wound: The True Price of America's Cold War Victory* (New York: Little, Brown, 2002), pp. 97, 399.
7. P. Mathias, "Skills and the Diffusion of Innovations from Britain in the Eighteenth Century," *Transactions of the Royal Historical Society,* 5th ser., 25 (1975): 93–113; D. J. Jeremy, "Damning the Flood: British Government Efforts to Check the Outflow of Technicians and Machinery, 1780–1843," *Business History Review,* 51 (1977): 1–34; and two early essays that would eventually become chapters in J. R. Harris's posthumous masterwork, *Industrial Espionage and Technology Transfer: Britain and France in the Eighteenth Century* (Brookfield, VT: Ashgate, 1998).
8. Philip Hanson, private correspondence (letter), January 13, 2005.
9. Dwight D. Eisenhower, "State of the Union Address" (1956), Docu-

ments for the Study of American History Website, http://www.ku.edu/carrie/docs/texts/dde1956.htm (retrieved on April 10, 2005).

10. Gordon Brook-Shepherd, *The Storm Birds: Soviet Post-War Defectors* (London: Weidenfeld and Nicolson, 1988), pp. 197–199; Christopher Andrew and Oleg Gordievsky, *KGB: The Inside Story of Its Foreign Operations from Lenin to Gorbachev* (London: Hodder and Stoughton, 1990), pp. 435, 480.

11. Philip Hanson, *Soviet Industrial Espionage: Some New Information*, RIIA papers (London: Chatham House, 1987), p. 18.

12. Gus W. Weiss, "Duping the Soviets: The Farewell Dossier," *Studies in Intelligence*, 39, no. 5. Available at http://www.cia.gov/csi/studies/96unclass/farewell.htm. Downloaded December 21, 2004. Eventually, the CIA would heed these and similar warnings and create section K under Stansfield Turner in the late 1970s, according to Peter Schweizer, *Victory: The Reagan Administration's Secret Strategy That Hastened the End of the Cold War* (New York: Atlantic, 1994), p. 47.

13. Weiss himself would later receive, in 1980, the Intelligence Medal of Merit "in recognition of his significant contributions to the Central intelligence Agency from 1972 to 1980 [for working] . . . closely with numerous Agency offices in the fields of international economics, civil technology, and technology transfer." Leebaert, *The Fifty-Year Wound*, p. 398.

14. Ibid., p. 396.

15. Harris A. Gilbert, JD, personal correspondence, January 14, 2005.

16. Leebaert, *The Fifty-Year Wound*, p. 399.

17. When President Carter ordered American companies to stop shipping spare computer parts to the USSR, productivity at the Kama River plant soon dropped radically. Ibid., pp. 395–396, 483. Gilles Kepel, *The War for Muslim Minds* (Cambridge: Harvard University Press, 2004), pp. 52–53.

18. Andrew and Gordievsky, *KGB: The Inside Story*, pp. 389–390, 393–395.

19. Schweizer, *Victory*, p. 110.

20. Jeremy Leggett, *The Carbon Wars: Global Warming and the End of the Oil Era* (New York: Routledge, 2001), pp. 63–64.

21. The first complete description of this enterprise appeared in 1983, supposedly written under an alias by Raymond Nart, the DST's Moscow station master at the time of the Farewell operation. See Henri Regnard, "L'URSS et l'information scientifique, technologique et tech-

nique," *Defense Nationale*, December 1983, pp. 107–121. Other descriptions appeared after Vetrov's execution (January 23, 1985), including reports in *Financial Times*, March 30, 1985, and *Le Monde*, April 2, 1985.

22. Cited in Gus W. Weiss Jr., "The Farewell Dossier: Strategic Deception and Economic Warfare in the Cold War" (unpublished essay, 2003), p. 2. These remarks were denoted "The New Brezhnev Approach" by the U.S. Department of Defense in Hearings before the House Committee on Banking and Currency (1974), p. 800. Leebaert, *The Fifty-Year Wound*, p. 512.

23. These details would later be fictionalized in John Le Carré's *The Russia House* (1989), in which Farewell's original contact, Jacques Prevost, a highly successful French corporate leader, and M. Ameil, the mole's first contact, are transformed into a single person, the unsuccessful English publisher Barley Scott Blair. Le Carré also appears to have substituted Soviet rocket scientist Sergei Korolev for a more realistic portrayal of Vladimir Vetrov. While it was too dangerous for Gordievsky to escape the Soviet Union with a copy of his KGB history, he later collaborated on a Western version with a British intelligence historian. See Andrew and Gordievsky, *KGB: The Inside Story*.

24. Ibid., pp. 389–390, 393–395.

25. Sergei Kostine, *Bonjour Farewell: La verite sur la taupe francaise du KGB* (Paris: Robert Laffont, 1997), p. 33.

26. Andrew and Gordievsky, *KGB: The Inside Story*, p. 515.

27. Kostine, *Bonjour Farewell*, pp. 73–74.

28. Hanson, *Soviet Industrial Espionage*, p. 21. Andrew and Gordievsky, *KGB: The Inside Story*, p. 423.

29. Ibid.

30. Ibid.

31. A British economist posted to Moscow at the time recalls the difficulty of contact between Russians and foreigners: "Any Russian who fraternised with you either had a license to do so or was unusually naive—if the latter, their fraternising with a foreign embassy official put them at risk. Like some people in Western embassies then and later, I had some Russian friends from earlier, non-governmental visits. This was awkward, but the independent-minded Russians concerned knew the score and too the risk of coming openly to dinner at my flat—which they knew would be recorded and might be used against them. None

suffered as a result. But in general it was understood that in the Embassy (except in a safe room) and in one's apartment, what was said was being recorded, so contacts were unavoidably constrained. The Embassy people who, in my experience, had the best contacts, were the military attaches. This may seem odd, but they developed some fellow-professional mutual understanding with their Russian official contacts. The wife of an Embassy friend of mine who was a keen rider, also developed a circle of local horsey friends that seemed to be almost outside politics. My job was economic reporting. I was expelled as part of a retaliation for a monumental exercise when 105 Soviet officials were expelled from London for 'activities incompatible with diplomatic status' on both sides. The real intelligence people in our Embassy were not touched. I didn't get back into the country until 1983—oddly enough, at a time of difficult E-W relations. A standard way for Embassy officials to get more relaxed, informal contacts, was to travel. On trains etc. it was easier to talk. Of course, visiting as a tourist or exchange scholar in those days also entailed being under surveillance." Philip Hanson, personal correspondence, January 13, 2005.

32. Kostine, *Bonjour Farewell*, p. 240.
33. Andrew and Gordievsky, *KGB: The Inside Story*, p. 489.
34. Kostine, *Bonjour Farewell*, p. 240.
35. Ibid., pp. 229, 237, 239.
36. Ibid., pp. 259–260.
37. Yves Bonnet, *Contre-Espionage: memoires d'un patron de la DST* (Paris: Calmann-Levy, 2000), p. 94. Later, Kostine would revise his theory and claim that Edward Lee Howard, an American CIA agent working for the KGB, had exposed Vetrov.
38. Brook-Shepherd, *The Storm Birds: Soviet Post-War Defectors*, p. 264. Kostine, *Bonjour Farewell*, p. 268.
39. Andrew and Gordievsky, *KGB: The Inside Story*, pp. xix–xxx. For details see Bonnet, *Contre-Espionage*.
40. This "conversation" between Weiss and Casey is a dramatic recreation based on substantive descriptions in the following sources: Norman A. Bailey, *The Strategic Plan That Won the Cold War: National Security Decision Directive 75*, 2nd ed. (MacLean, VA: Potomac Foundation, 1999), p. 18; Leebaert, *The Fifty-Year Wound*, p. 527; Thomas C. Reed,

At the Abyss: An Insider's History of the Cold War (New York: Random House, 2004), pp. 266–270; Thomas C. Reed, personal correspondence, January 26, 2005; William Safire, "The Farewell Dossier," *New York Times,* February 2, 2004; Gus W. Weiss Jr., "Cold War Reminiscences: Super Computer Games," *Intelligencer: Journal of U.S. Intelligence Studies* (Winter/Spring 2003): 57–60; Weiss, "Duping the Soviets: The Farewell Dossier"; Gus W. Weiss Jr., "The Farewell Dossier: Strategic Deception and Economic Warfare in the Cold War" (unpublished, 2003), pp. 7–11.

41. Schweizer, *Victory,* p. 62.
42. Leebaert, *The Fifty-Year Wound,* p. 527.
43. Ibid., pp. 43, 71, 74.
44. Ibid., p. 83.
45. Ibid., p. 110; see also pp. 82–83.
46. Ibid.
47. Safire, "The Farewell Dossier."
48. Reed, *At the Abyss,* pp. 268–269.
49. Ibid., p. 268.
50. Leggett, *The Carbon Wars,* p. 67.
51. Leebaert, *The Fifty-Year Wound,* p. 509.
52. Schweizer, *Victory,* p. 83; Bailey, *The Strategic Plan That Won the Cold War,* p. 9.
53. Leggett, *The Carbon Wars,* p. 64.

9. Cell Phones and E-Waste

1. Betty Fishbein, *Waste in the Wireless World* (New York: Inform, 2001), pp. 22, 27.
2. Silicon Valley Toxics Coalition, *Poison PCs and Toxic TVs: California's Biggest Environmental Crisis that You've Never Heard Of* (San Jose: 2001), p. 8.
3. Fishbein, *Waste in the Wireless World,* p. 22.
4. Charles Babbage, *On the Economy of Machines and Manufacturers* (1832), available online from Project Gutenberg: http://www .gutenberg.net/etext03/cnmmm10.txt: "Machinery for producing any commodity in great demand seldom actually wears out; new improvements, by which the same operations can be executed either more

quickly or better, generally superseding it long before that period arrives: indeed, to make such an improved machine profitable, it is usually reckoned that in five years it ought to have paid itself, and in ten to be superseded by a better."

5. Fishbein, *Waste in the Wireless World*, pp. 3, 20, 24; Carl H. Marcussen, "Mobile Phones, WAP and the Internet," Research Center of Bornholm Denmark, October 22, 2000.

6. Fishbein, *Waste in the Wireless World*, p. 21. Eric Most, *Calling All Cell Phones: Collection, Reuse and Recycling Programs in the US* (New York: Inform, 2003), p. 2.

7. Silicon Valley Toxics Coalition, *Poison PCs and Toxic TVs*, p. 1.

8. Mary Jo Leddy, *Radical Gratitude* (Maryknoll, NY: Orbis Books, 2003), pp. 20, 21. Colin Campbell, *The Romantic Ethic and the Spirit of Modern Consumerism* (Oxford: Basil Blackwell, 1987), p. 37.

9. Colin Campbell, "The Desire for the New: Its Nature and Social Location as Presented in Theories of Fashion and Modern Consumerism," in Roger Silverstone and Eric Hirsch, eds., *Consuming Technologies: Media and Information in Domestic Spaces* (London: Routledge, 1992), pp. 53–54.

10. For a useful summary of how sociology has tried to come to terms with society's relationships to technology, including the current dominant theory of "domestication" which informs his own work, see Rich Ling, *The Mobile Connection: The Cell Phone's Impact on Society* (San Francisco: Kaufmann, 2004), pp. 21–33. A second useful discussion of the history of such theories and an insightful analysis of domestication appears in Campbell, *The Romantic Ethic*, pp. 36–57. Roger Silverstone, Eric Hirsch, and David Morley also provide a useful point-by-point summary of "domestication" in "Information and Communication Technologies and the Moral Economy of the Household," in Roger Silverstone and Eric Hirsch, eds., *Consuming Technologies: Media and Information in Domestic Spaces* (London: Routledge, 1992), pp. 15–31, while James E. Katz's essay, "Do Machines Become Us?" in James E. Katz, ed., *Machines That Become Us* (New Brunswick, NJ: Transaction Publishers, 2003), pp. 15–25, explains what has been described as the American "apparatgeist" approach, an adaptation of "domestication." See also Campbell, "The Desire for the New," pp. 55–56.

11. In *The Romantic Ethic*, p. 60, Campbell points out that this view is

suggested by Edward O. Laumann and James S. House, "Living Room Styles and Social Attributes: The Patterning of Material Artifacts in a Modern Urban Community," in H. H. Kassarjaian and T. S. Robertson, eds., *Perspectives in Consumer Behavior* (Glenview, IL: Scott Foresman, 1973), pp. 430–440.

12. Campbell, "The Desire for the New," p. 56.
13. Ibid., pp. 56–57.
14. Cass R. Sunstein, *Why Societies Need Dissent* (Cambridge: Harvard University Press, 2003), pp. 10–11.
15. Ling, *Mobile Connection,* p. 6, n.8.
16. Ibid., pp. 11, 14–15, 86.
17. Ibid., p. 96.
18. Ibid., p. 104.
19. Ibid., pp. 97, 15.
20. Paul Levinson, *Cellphone: The Story of the World's Most Mobile Medium and How It Has Transformed Everything* (New York: Palgrave Macmillan, 2004), p. 127. Patricia Riedman, "U.S. Patiently Awaits Wireless Texting That's Soaring Overseas," *Advertising Age,* 73 (April 2002): 6. Lisa W. Foderaro, "Young Cell Users Rack Up Debt, A Message at a Time," *New York Times,* January 9, 2005.
21. Ibid.
22. Campbell, "The Desire for the New," p. 54.
23. Ling, *Mobile Connection,* pp. 58, 73.
24. Ibid., pp. 62, 76.
25. Ibid.
26. Ibid., p. 81.
27. Ibid., p. 67.
28. Ling, *Mobile Connection,* pp. 80, 208, n.28.
29. David Pogue, "Wristwear: Zap, Snap, or Zero In," *New York Times,* May 24, 2001, p. G1. Julie Dunn, "Looking In on a Loved One," *New York Times,* February 17, 2001.
30. Fishbein, *Waste in the Wireless World,* pp. 5, 27.
31. Susan L. Cutter, *Living with Risk: The Geography of Technological Hazards* (New York: Hodder and Stoughten, 1993); Lloyd J. Dumas, *Lethal Arrogance: Human Fallibility and Dangerous Technologies* (New York: St. Martin's, 1999); H. W. Lewis, *Technological Risk* (New York: Norton, 1990), p. 26; Stephanie Mills, ed., *Turning Away from Technology: A New Vision for the Twenty-First Century* (San Francisco: Sierra Club,

1997); Thomas P. Hughes, *Human-Built World* (Chicago: University of Chicago Press, 2004); P. Aarne Vesilind and Alastair S. Gunn, eds., *Engineering, Ethics, and the Environment* (Cambridge: Cambridge University Press, 1998).

32. George Carlo and Martin Schram, *Cell Phones: Invisible Hazards in the Wireless Age* (New York: Carroll and Graf, 2001), pp. 149–179.

33. Blaine Harden, "The Dirt in the New Machine," *New York Times*, August 13, 2001, p. SM36.

34. Basel Action Network, Silicon Valley Toxics Coalition, *Exporting Harm: The High Tech Trashing of Asia* (Seattle: 2002).

35. Richard Black, "E-waste Rules Still Being Flouted," BBCNews online, March 19, 2004: http://news.bbc.co.uk.2/hi/science/nature/3549763 .stm. Accessed April 22, 2004.

36. Greg Pearson and A. Thomas Young, eds., *Technically Speaking: Why All Americans Need to Know More about Technology* (Washington, DC: National Academy Press, 2002), pp. 1–2.

37. *Standards for Technological Literacy: Content for the Study of Technology* (Reston, VA: International Technology Education Association, 2000).

Acknowledgments

I hope this will be the kind of book I wanted to write as a young man. It's for everything and for everyone I love: for my father, who had unrealized dreams and who first mentioned planned obsolescence to me; for the woman who loves me and insisted that I complete it; for our young sons, who may need to describe me to their own children some day, and for our newborn, whoever he may become.

I am an unfashionable Canadian who writes admiringly about America's excesses and success. This is because America accepted me for what I was, *gave* me the expensive education I chose, and then provided me with opportunities that still do not exist in my own country. Canada's greatness lies in the raw, physical beauty of its magnificent landscape. America too is beautiful, but its greatness rests in the generosity of its people and in their constantly renewing struggle to realize a flawed but wonderfully extravagant dream. Foreigners like me, who recognize the struggle and the dream, are doomed to a lifelong discontent with their own culture. Even though there are many things wrong with America and even though the dream has often been perverted and misused, this is still true. So, for my own discontent, America, thank you.

Special thanks to Jay Martin, and to my other teachers: Jackson I. Cope, Michael Dobrovolsky, Stanley Fish, David James, Donna Landry, Arch MacKenzie, Jean-Pierre Paillet, Ian Pringle, Nancy Vickers, and Rosemary Wood. Thanks also to Gary Cross, Jeffrey Meikle, Michael O'Malley, and in particular to Susan Strasser, for reading proposals and making suggestions in the early stages. Thanks to Tim Sullivan of Princeton University Press, Jenny Fyffe of Raincoast Press, and literary agent extraordinaire, John W. Wright, for practical suggestions about contracts and publishing. More thanks to all those who generously assisted in individual points of research, but especially to Emily Howie at the Library of Congress, Joseph Ditta at the New York Historical Society, and Michael Brian Schiffer at the University of Arizona.

My email contacts deserve special thanks since they freely answered annoying questions from an unaffiliated scholar. These include Glenn Adamson, Romaine Ahlstrom, Jean Ashton, Elizabeth Bancroft, Rosalyn Baxandall, Sharon Beder, Richard Black, Janet Brodie, Pam Brooks, Colin Campbell, Eric Chazen, Sally Clarke, Catherine Cookson, Alison Curtis, Sara Davis, Helga Dittmar, Maurice Eisenstein, Nichols Fox, Julia Franco, David Gartman, Harris Gilbert, Tom Gilbert, Ann Graham, Barbara Grimm, Scott Groller, Judy Growe, Jeanne Hammond, Philip Hanson, Margarite Holloway, Maryfran Johnson, Paul M. Johnson, Barbara M. Kelly, Greg Key, Judy Kornfield, Dan Lewis, P. J. Lenz, Steve McCoy, Mary McLuhan, Jeffrey Meikle, Rose Ann Miller, Robin Nagle, Henry Petroski, Cris Prystay, Thomas C. Reed, David Robarge, Heather Rogers, Elizabeth Royte, Michael Ruse, Janice Rutherford, Tom Savini, Roger Semmens, Fred Shapiro, Juliet Schor, Will Slade, J. Philip Steadman, Andrea Tone, Peggy Vezina, Laurie

Weisman, Eric Williams, Rosalind Williams, Joyce Wong, Tanya Zanish-Belcher, and Sharon Zukin. Let me also offer a special thanks to the friends, relatives, and colleagues of Gus W. Weiss Jr., who risked leaving the shadows long enough to communicate their memories and concerns clearly to me. (Thank you, friends—there is still a bit more to do.)

Finally, my sincerest thanks to Michael G. Fisher and Susan Wallace Boehmer of Harvard University Press, who always set the bar very high and told me "why" whenever I failed to clear it.

Index